FISH SMOKING AND DRYING

The Effect of Smoking and Drying on the Nutritional Properties of Fish

FISH SMOKING AND DRYING

The Effect of Smoking and Drying on the Nutritional Properties of Fish

Edited by

J. R. BURT

Torry Research Station, Aberdeen, UK

ELSEVIER APPLIED SCIENCE
LONDON and NEW YORK

ELSEVIER SCIENCE PUBLISHERS LTD
Crown House, Linton Road, Barking, Essex IG11 8JU, England

Sole distributor in the USA and Canada
ELSEVIER SCIENCE PUBLISHING CO., INC.
52 Vanderbilt Avenue, New York, NY 10017, USA

WITH 28 TABLES AND 6 ILLUSTRATIONS

British Library Cataloguing in Publication Data

Fish smoking and drying.
1. Food. Fish. Nutritional values. Effects of curing
I. Burt, J. R.
641.3′92

Library of Congress Cataloging in Publication Data

Fish smoking and drying.

1. Fish, Smoked. 2. Fish, Dried. I. Burt, J. R.
TX556.5.F52 1989 664′.946 88-16550

ISBN 1-85166-247-2

Printed in Great Britain at the University Press, Cambridge.

PREFACE

Reductions in the nutritional values of foodstuffs caused by the methods used to prepare, cook, process or store them have long been of interest and concern. Over the years, various aspects of this topic have been reviewed but it is over 25 years since the effects of drying and smoking on the nutritional value of fish have been reviewed in any detail (Cutting, 1962). Considerations such as these are of particular concern in developing countries, primarily those in the tropics, where wastage and shortages of food are unfortunately only too common. Many traditional products in these parts of the world have been developed, often centuries ago, so that they are stable and have storage properties that enable them to be distributed and stored in the absence of sophisticated refrigeration systems. The preservation methods used often rely on relatively harsh processing conditions and it is appropriate to ask the question, 'Do the techniques used destroy nutrients?' Having said that, however, one must not lose sight of the even more considerable losses which occur deliberately (such as when lower value species in a by-catch are returned to the sea) or adventitiously (such as through the action of pests or through the normal spoilage processes).

The subject of this book covers what might be described as some of the oldest methods of preserving fish. Their use is not restricted to the tropics and in the following chapters products from temperate regions are given similar consideration. The conditions of processing range from the relatively mild to the quite harsh in terms of time and temperature and the products are almost entirely those derived from true fish destined for human consumption. Occasionally other foods of marine origin, such as those based on seaweeds, molluscs or crustaceans, are discussed but the extensive literature on fish meal has been largely excluded. It has, however, proved necessary in some instances to cite certain studies on fish meal in order to throw light on a particular point but it must be emphasised that the studies cited may not be relevant to the generality of fish meals that are found in modern commercial trade. Dried fish of very low water activity can be regarded for some purposes as unground fish meal but there are some fundamental differences. Whole dried fish usually has fat just under the skin while in fish meal the fat is in and on the surface of every dried particle.

The composition of fish varies considerably even when samples are taken from a batch which has been caught on a particular ground at a particular time; and product composition changes with processing parameters. Standard quantities such as those used by nutritionists and dieticians are only indicative; comparisons of levels of constituents in different batches could lead to misleading conclusions. It would be even more misleading if the average values pertaining to raw materials and to products were taken from nutritional tables and any differences attributed to the respective conditions of processing. Attempts have been made in the study reported here to consider mainly the literature covering experiments where the effects ascribed to processing relate to samples derived from the same batch.

It was felt that the effects of storage subsequent to processing should be considered as well and in this regard water activity (the effective, as distinct from the actual, water concentration) would appear to be of prime importance in controlling spoilage due to bacteria or fungi in foods held unrefrigerated. The first chapter deals extensively with water activity and how its effect on shelf life can be assessed whilst the last chapter gives several water activity values which can be used for predictive purposes, albeit cautiously.

It is appropriate at this stage to reiterate some words of caution relating to the methods of analysis which have been used over the years. Different methods for analysing any one component do exist and results can vary widely; furthermore, analytical techniques have progressively improved, hopefully becoming more precise and more specific. It is therefore essential to be sure that comparisons only of like with like are made before too firm conclusions are drawn. Of particular relevance in this context are the various chemical and biological tests which have been proposed and devised for giving indications of the nutritional quality of proteins. For instance, with the pepsin digestibility test, where early physiologists tried to simulate true digestion and later ones attempted to speed up the assay, the values obtained are very technique dependent. The particular variant of the technique needs to be taken into account and results from *in vitro* experiments must not be regarded as absolute but only comparative within each particular trial (Lovern, 1965).

Another warning which is applicable to all literature surveys, and not just to the types of study reviewed here, relates to the necessity wherever possible to refer back to original publications. Errors can unfortunately arise and misquotations appear in reviews and even in abstracts; these can be repeated in good faith and occasionally become enshrined in the literature.

This book consists of contributions to the subject made by different members of a working group set up under the joint aegis of the International Union of Food Science and Technology and the International Union of Nutritional Sciences. The remit of this working group had arisen out of a concern which had been expressed by Dr J. Hulse, a former President of IUFoST, and Dr N. Scrimshaw about the effects the consumption of smoked and dried fish might have on the gut pathology of peoples in developing countries. Most of the work was done by correspondence but, thanks to the generosity of these two international unions, the working group was able to meet in Dublin under the auspices of the Sixth International Congress of Food Science and Technology.

The first six chapters relate more strictly to the topic described by the extended title of the book, each dealing with designated aspects of the topic. Not only are changes in desirable nutrients such as proteins, lipids and vitamins covered but the introduction into the foods of what might be regarded as undesirable constituents and reaction products is also considered. Chapters 7 and 8 are more concerned with the ways in which various dried and smoked products are prepared. Finally, Chapter 9 gives in tabular form details of the composition and properties of products arranged primarily according to their country of origin.

Thanks are due to the members of the working group and to certain of their colleagues who assisted in the collection of data and in the preparation of the various chapters. The group is also indebted to Mr David James, FAO, Rome, who contributed extensively to the discussions in Dublin and who, with the assistance of participants in the FAO networks of cooperation in fish technology in Asia and Africa, provided much of the data on traditional, regional dried and smoked products and to Dr Preeya Vilbulsresth, Karetsart University, Bangkok, for the information on Thai products. Finally, the help given by Mrs M. Tevendale in typing the various drafts of the chapters is gratefully acknowledged.

The following are members of the working group, some being associated with it for only part of its life:

Dr E. G. Bligh, *Technical University of Nova Scotia, 1360 Barrington Street, P.O. Box 1000, Halifax, Nova Scotia B3J 2X4, Canada.*

Dr M. Caurie, *The Federal Polytechnic, PMB 50, Ilaro, Ogun State, Nigeria.*

Dr P. E. Doe, *Department of Mechanical Engineering, University of Tasmania, Box 252C GPO, Hobart, Tasmania, Australia 7001.*

Dr W. Edwardson, *International Development Research Centre, University of Alberta Campus, Edmonton T6G 2G6 Canada.*

Dr M. S. Ismail, *Universiti Pertanian Malaysia, Serdang, Selangor, Malaysia.*

Professor T. Motohiro, *Kagoshima University, Faculty of Fisheries, Laboratory of Food Quality Control and Analysis, Shimoarata, Kagoshima, Japan.*

Dr J. Olley, *CSIRO, Division of Fisheries, 'Stowell', Stowell Avenue, Hobart, Tasmania 7004, Australia.*

Dr J. Opstvedt, *Norwegian Herring Oil & Meal Industry Research Institute, 5033 Fyllingsdalen, Bergen, Norway.*

Professor B. S. Pan, *National Taiwan College of Marine Science & Technology, Keelung, Taiwan, China.*

Dr R. G. Poulter, *Overseas Development Natural Resources Institute, 56–62 Gray's Inn Road, London WC1X 8LU, UK.*

Professor Z. E. Sikorski, *Department of Food Preservation and Technical Microbiology, Technical University Politechnika Gdanska, 80–952 Gdansk, Poland.*

Professor E. C. Sison, *College of Agriculture, University of the Philippines at Los Banos, College, Laguna 3720, Philippines.*

REFERENCES

Cutting, C. L. (1962). The influence of drying, salting and smoking on the nutritive value of fish. In: *Fish in Nutrition*, E. Heen and R. Kreuzer (Eds), Fishing News (Books), London, pp. 161–79.

Lovern, J. A. (1965). Some analytical problems in the analysis of fish and fish products. *J. Ass. Off. Agric. Chem.* **48**, 60–8.

Dr J. R. Burt
Chairman, Joint IUFoST/IUNS Working Group
Torry Research Station
PO Box 31, 135 Abbey Road
Aberdeen AB9 8DG, United Kingdom

CONTENTS

LIST OF CONTRIBUTORS

E. G. Bligh

Technical University of Nova Scotia, 1360 Barrington Street, P.O. Box 1000, Halifax, Nova Scotia B3J 2X4, Canada.

J. R. Burt

Torry Research Station, P.O. Box 31, 135 Abbey Road, Aberdeen AB9 8DG, UK.

P. E. Doe

Department of Mechanical Engineering, University of Tasmania, Box 252C GPO, Hobart, Tasmania, Australia 7001.

E. S. Heruwati

Institute of Fishery Technology, JL Karel S Tubun, Pos Palmerah, Jakarta, Indonesia.

T. Motohiro

Kagoshima University, Faculty of Fisheries, Laboratory of Food Quality Control and Analysis, Shimoarata, Kagoshima, Japan.

J. Olley

CSIRO, Division of Fisheries, 'Stowell', Stowell Avenue, Hobart, Tasmania 7004, Australia.

J. Opstvedt

Norwegian Herring Oil & Meal Industry Research Institute, 5033 Fyllingsdalen, Bergen, Norway.

B. S. Pan

National Taiwan College of Marine Science & Technology, Keelung, Taiwan, China.

R. G. Poulter

Overseas Development Natural Resources Institute, 56–62 Gray's Inn Road, London WC1X 8LU, UK.

S. J. Shaw

Technical University of Nova Scotia, 1360 Barrington Street, P.O. Box 1000, Halifax, Nova Scotia B3J 2X4, Canada.

Z. E. Sikorski

Department of Food Preservation and Technical Microbiology, Technical University Politechnika Gdanska, 80–952 Gdansk, Poland.

A. D. Woyewoda

Technical University of Nova Scotia, 1360 Barrington Street, P.O. Box 1000, Halifax, Nova Scotia B3J 2X4, Canada.

Chapter 1

THE INFLUENCE OF DRYING AND SMOKING ON THE NUTRITIONAL PROPERTIES OF FISH: AN INTRODUCTORY OVERVIEW

J. OLLEY
CSIRO, Division of Fisheries, Tasmania, Australia

P. E. DOE
University of Tasmania, Hobart, Australia

E. S. HERUWATI
Institute of Fishery Technology, Jakarta, Indonesia

INTRODUCTION

Smoked and dried fish is a traditional part of the diet of a large section of the world's population. In recent years the twin effects of increasing population and increased pressure on the established fisheries has meant that any diminution of nutritional quality of smoked and dried fish or any fall in yield can result in severe protein deficiencies for whole communities (FAO, 1981; James, 1984).

The IUFoST/IUNS working group on the influence of smoking and drying on the nutritional and functional properties of fish, recognised the need for a compendium of information on the yield from different production methods, as well as the effects on the nutritional and functional properties.

Partly as a result of the Working Group's initiatives, a research team supported by the Australian Centre for International Agricultural Research and the Agency for Agricultural Research and Development (Indonesia) undertook a 3 year study into the spoilage of fresh, cured and dried tropical fish by bacteria, moulds and insects.

The Workshop's report therefore begins with a review of the composition of fish, the various ways in which spoilage can occur, the factors which affect the rate of spoilage and the effects of processing and spoilage on the nutritional quality of the product with particular reference to water activity.

The sequence of changes in fish quality due to a number of causes, such as increases in bacterial numbers in the wet fish, mould growth due to slow drying, and/or lengthy storage accompanied by a succession of flies, their larvae and beetles, is well described in James (1983).

It is widely known that reducing the water activity (Aw) will result in a reduction of microbial activity; at low water activities, however, other effects such as rancidity become important (Troller and Christian, 1978). An important aim in the reduction of nutritional losses and wastage is producing a cured, smoked or dried fish product with a value of water activity as low as possible without exceeding desired levels of salt, or being so dry as to be unpalatable or irreversibly intransigent to reconstitution. The amount of salt and moisture needed to give a desired water activity can be calculated (Doe *et al.*, 1982).

WATER ACTIVITY

The basic concepts of water activity and water sorption isotherms of foods are described by Troller and Christian (1978) and Lupin (1986). Water activity, Aw, is a measure of the water available in a system for biochemical and chemical reactions, and for microbial activity (FAO, 1981).

Aw is defined as:

$$\mathrm{Aw} = p/p_0$$

where p is the vapour pressure of the product, and p_0 is the vapour pressure of pure water at the same temperature.

The water activity of a foodstuff can be measured by enclosing a sample in a sealed container so that it reaches equilibrium with the surrounding air. This equilibrium relative humidity of the air is measured using some form of hygrometer. The water activity is equated to the measured relative humidity expressed as a fraction rather than as a percentage.

The relationships between the equilibrium moisture content and relative humidity at different temperatures (sorption isotherms) are slightly different depending on whether the food is drying (desorption isotherm) or rehydrating (adsorption isotherm) as it approaches equilibrium moisture content (Moschair *et al.*, 1984). Desorbed foods are less stable than foods with the same water activity prepared by adsorption (Sperber, 1983).

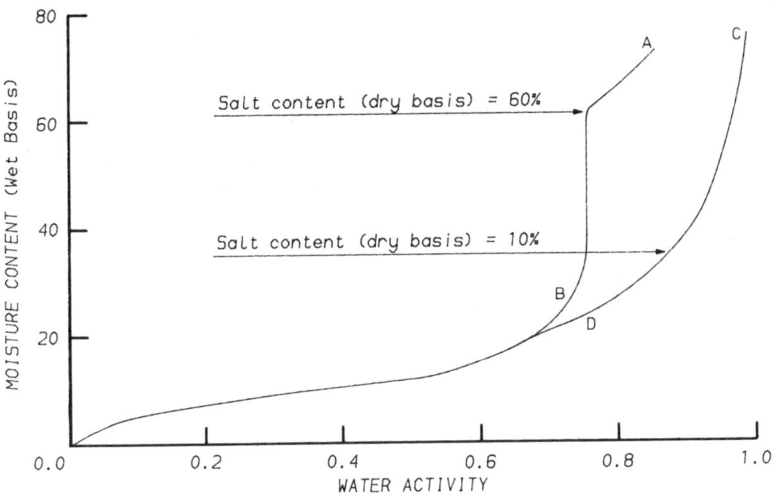

FIG. 1. Sorption isotherms for salted (A–B) and unsalted (C–D) cod.

For ideal solutions (real solutions at low concentrations), water activity can be calculated from the expression:

$$Aw = n_1/(n_1 + n_2)$$

where n_1 is the number of moles of solvent, and n_2 is the number of moles of the solute.

For cured fish the relationship becomes more complex due to the interactions between water and fish flesh and the high concentrations involved. Drying the fish lowers n_1 and the addition of salt increases n_2; the result is a decrease in the water activity of the product. If, however, the relative humidity of the air rises above that of the equilibrium relative humidity of the product, the surface will tend to absorb water. Conversely, a lower ambient relative humidity will result in a loss of moisture from the fish.

The effect of salt on the drying behaviour and water activity of fish is illustrated in Fig. 1. Salted fish dries along the path A–B in Fig. 1. With an initial moisture content (after salting) of around 70% (wet basis), only another 5% of moisture need be removed by drying before the water activity of the fish is reduced to 0·75. On the other hand, with unsalted fish, the initial moisture content is higher (about 75% wet basis) and must be reduced to near 20% moisture content to achieve the same drop in Aw (path C–D). A high salt fish, or a high fat fish, may appear to be relatively

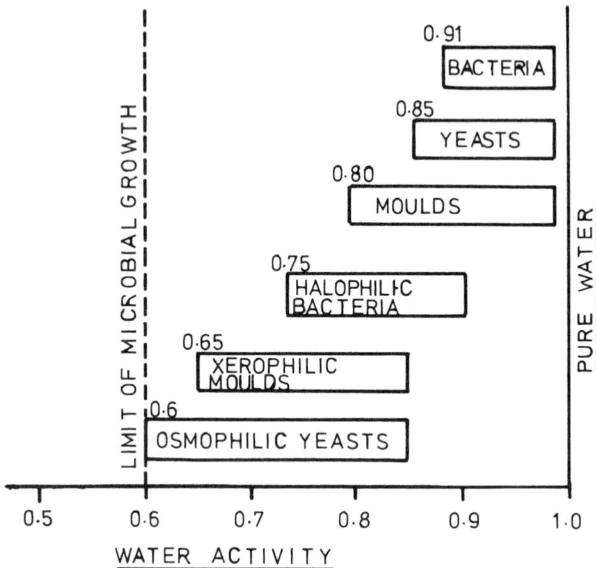

FIG. 2. Growth ranges of microorganisms with respect to water activities.

moist and pliable and can have a high moisture content yet can have the same water activity as a rock-hard unsalted, lean dried fish. Salt uptake is more rapid in stale fish, but salt decreases the rate of drying (Clucas and Sutcliffe, 1981; Aima *et al.*, 1984).

All microorganisms have an absolute requirement for water in order to grow. The range and minimum levels of water for the growth of various organisms is shown in Fig. 2, taken from FAO (1981). Lupin (1986) has completed a more comprehensive list of minimum Aw values to which the list of common moulds on fish in Indonesia must be added (Anggawati *et al.*, 1985/86).

Polypaecilum pisce (Pitt and Hocking, 1985), which has an optimum growth at 30°C between Aw values 0·96 and 0·90 on media containing salt, has favourable conditions for growth on Indonesian dried fish. The occurrence of the 'dun' mould *Wallemia sebi* (*Sporendonema*), traditionally regarded as the principal cause of spoilage of salt fish, is rare in tropical regions – probably due to higher ambient storage temperature. Growth of *W. sebi* is poor at 30°C and almost completely inhibited at 34°C especially in the presence of salt.

The moulds present on a food are not necessarily directly related to the

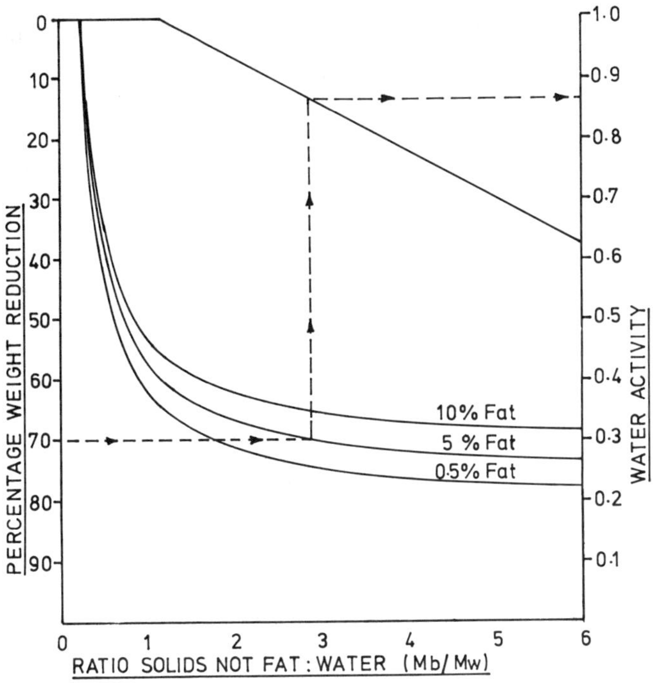

FIG. 3. Graphical estimation of Aw in fish drying from percentage of weight change for lean, medium fat and fatty fish (after Lupin, 1986).

measured water activity of the food at a particular time. It may be that the moulds grew at a time when the water activity was at some different (higher) value and are now inactive. If this is the case, the proportions of the different moulds present on a food sample reflect the temperature and water activity history of the sample. For the same reason, the bacterial count in a food is related to the temperature history of the food. Conversely, the moulds and bacteria found in a fish sample are those most likely to predominate if the sample becomes rehydrated.

Doe *et al.* (1982) have shown that the sorption behaviour of dried salted cod (a very low fat fish) can be predicted from the physical properties of sodium chloride solutions and the sorption characteristics of the fish muscle. A similar result has been found for a medium fat tropical fish (Curran and Poulter, 1983). The isohalic sorption isotherms are derived using the Ross (1975) equation. At any particular water activity,

$$\mathrm{Aw} = \mathrm{Aw}_n . \mathrm{Aw}_0$$

where Aw_n is the water activity of the sodium chloride solution and Aw_0 is the water activity of the unsalted fish muscle.

Cod, *Gadus morrhua*, from temperate waters and lisa, *Xenomugil thoburni*, from tropical waters have been found to have very similar Aw_0 sorption isotherms (Curran and Poulter, 1983; Curran, 1984). However, crustaceans and cephalopods were found to have a higher moisture content than cod for a given water activity in the critical range of Aw from 0·7 to 0·75 (Curran, 1984).

There are various devices for measuring water activity (Troller and Christian, 1978), but many of the instantaneous devices have not lent themselves to use in the tropics where corrosion and other factors can be a problem (Curran, 1984). An alternative to the direct measurement of water activity is the measurement of the amount of salt, water and fat-free solids. The water activity can then be calculated using the method of Doe *et al.* (1983). Curran (1984) has compiled a table of the predicted mould-free shelf lives of Malaysian cured fishery products based on calculated values of Aw. However, they cannot predict the shelf life of redried stored products as the microbial attack which may occur between drying and redrying is not reversible.

Water activity can also be estimated from the weight reduction during drying (Fig. 3).

COMPOSITION OF FISH

The main nutritional components of fish are summarised by Love (1982). Fish flesh contains most of the minerals necessary for a balanced diet. In particular, fish is high in iodine. When fish bones are eaten, fish affords a valuable source of calcium and phosphorus (Nettleton, 1985). In general the water and fat content of fish together add up to about 80% of the tissue. This means that in lean fish such as cod, the water content is about 80%, while in fatty fish such as mackerel it may be as low as 50% depending on the season. Generally the protein content is about 18% but varies markedly between species depending on age and condition (Love, 1970). The main fibrillar proteins of fish are actin and myosin, responsible for the contraction and relaxation of muscles.

The complex fats which are components of cell membranes are phospholipids and cholesterol, while the depot fats used for energy are usually triglycerides. The depot fat in fish does not contribute to the water activity of the product and the Aw must be calculated on a water:fat-free solids, salt:fat-free solids basis (Doe *et al.*, 1983).

All the vitamins recognised as necessary for human nutrition are found in fish, but their distribution in the various tissues is very uneven. Fish oils are the richest known sources of vitamins A and D (Cruickshank, 1962).

Tocopherol or vitamin E, the natural antioxidant in fish flesh, is of importance in that it protects the polyunsaturated fatty acids in fish flesh, now thought to be important in the prevention of heart disease (Dyerberg and Jørgenson, 1982).

Chou and Labuza (1974) found that antioxidants tended to be more effective for a given Aw in absorption systems; that is a dried product which is taking up water. The hysteresis effect on Aw is well described in Troller and Christian (1978).

THERMAL PRINCIPLES OF THE SPOILAGE PROCESS

Microbiological and Chemical Changes

Spoilage begins as soon as the fish dies. The mechanism of spoilage is complex, but is usually characterised as being due to enzyme action (autolysis), change in redox potential, nucleotide breakdown, bacterial action, chemical changes such as oxidation leading to rancidity and physical damage including insect attack.

The key environmental factors affecting the reaction have been reviewed by Karel (1984). Enzymes present in the living fish remain active after the death of the fish. Some of these, in particular the enzymes contained in the gut, are very effective in breaking down fish flesh (Gildberg, 1982).

Bacteria are always present on the surface and in the gut of fish, in fish boxes and on salt and ice. These organisms enter the fish through the body openings and the skin. Good gutting can reduce the initial bacterial load. Poor gutting can spread bacteria all over the flesh and thus greatly increase the initial load.

Bacterial and enzymic spoilage are most effectively curtailed by a reduction in temperature. Enzymic reactions such as proteolysis and lipolysis respond to temperature roughly according to the Arrhenius equation (Labuza, 1984), but recently it has been noted that nucleotide breakdown (Ohta and Hirahara, 1977) and the growth of bacteria, yeasts and moulds vary more closely with temperature according to the relationship;

$$\sqrt{r} = b(T - T_{\min}) \qquad (1)$$

where r is the rate or reciprocal of time to reach a specified spoilage level, b is the slope of the regression line, T is temperature, and T_{min} is the intercept of the growth curve on the temperature axis; T_{min} corresponds to an infinitely large activation energy in the Arrhenius equation.

Equation (1) is a particular case of a more general equation for biotic growth first proposed by Belehrádek (1935).

This equation, discussed by Ratkowsky *et al.* (1982), and Pooni and Mead (1984), has been extended to cover the higher temperatures where microbial organisms and their enzymes are destroyed by heat (Ratkowsky *et al.*, 1983). Thus

$$\sqrt{r} = b(T - T_{min})\{1 - \exp[c(T - T_{max})]\} \tag{2}$$

The relative growth rate can be calculated directly from the expression

$$r_{T_1}/r_{T_2} = \{(T_1 - T_{min})/(T_2 - T_{min})\}^2$$

in which case the value of b need not be known.

For psychrotrophic spoilage the temperature T_{min} is in the range −10°C to 0°C, while for mesophiles the temperature range is 0°C to 10°C (Olley, 1983).

It is interesting to note that nucleotide breakdown in spoiling fish is thought to be more due to bacteria than was originally supposed (Sumner *et al.*, 1984), and the minimum temperature for this reaction on chilled fish is in the psychrotrophic range (Ohta and Hirahara, 1977). There is evidence that species of bacteria found on tropical fish can grow over a much wider range of temperatures than those associated with fish from temperate waters (J. L. Sumner and E. Gorczyca, unpublished data).

Hypoxanthine, one of the final products of nucleotide breakdown, is often used as a measure of the degree of spoilage. Fish which spoil through hypoxanthine production will have a shorter shelf life than those in which the nucleotides are dephosphorylated only to inosine. This explains why some tropical fish species spoil faster than others when stored under similar conditions; those which produce hypoxanthine will be rejected because of flavour sooner than those species which do not produce hypoxanthine, which will be rejected when levels of bacteria increase to unacceptable limits (Bremner *et al.*, 1984/85).

Other compounds used for the determination of specific spoilage levels such as trimethylamine (TMA) and total volatile base (TVB) are listed by Olley and Ratkowsky (1973). There are also instruments, such as the Torrymeter, which use electrical properties of the fish as measures of freshness (Hoffman, 1981).

Visual and taste panel assessments, however, remain the fundamental methods of determining freshness and acceptability of fish products (Branch and Vail, 1985; Anggawati *et al.*, 1985).

The temperature history of fish from the moment of death can be used as a means of estimating the remaining shelf life of the product and instruments measuring temperature and programmed with the square root relationship for psychrotrophs or mesophiles are now available (Owen and Nesbitt, 1984). Enzymic time–temperature monitor strips (Olley and Lisac, 1985) can also be prepared.

Unless otherwise limited by reasons of cost, availability or the dictates of a particular product, the first method of ensuring the maximum freshness of fish is to reduce the temperature by icing or refrigeration.

Autolytic Changes

The viscera of fish are sources of very active proteolytic enzymes, as are the anaerobic bacteria within the gut. These enzymes are very effective in breaking down fish flesh and may leak out from a seemingly intact digestive tract (Gildberg, 1982). Gildberg has made a most thorough study of the effect of visceral autolysis of fish tissues. He has examined the effect of digestive proteases from temperate and tropical fish at pH 3 and pH 9. The proteases are more active at the acid pH than the alkaline. Acid proteases are more active than alkaline ones at lower temperatures. The acid proteases have a relative response to temperature similar to that of psychrotrophic bacteria, while the relative response of the alkaline proteases more resembles that of mesophilic bacteria.

EFFECTS OF TEMPERATURE, pH, AND WATER ACTIVITY

Relative Importance

During the drying process and when fish are stored at constant relative humidity, the effects of decreased water activity become increasingly important to the microbial stability of the product. Doe (1983) found that the square root relationship of Ratkowsky *et al.* (1983) (eqn 2) fitted the data of Ayerst (1969) on mould germination and growth at water activities ranging from 0·75 to 0·90. The same approach has been taken with the data on the growth of *Clostridium botulinum* Type B (Troller and Christian, 1978), shown in Table 1.

At pH 6 the rate of growth of *C. botulinum* Type B was halved by either reducing the temperature from 40°C to 30°C or reducing the water activity

TABLE 1

RECIPROCAL OF TIME TO VISIBLE GROWTH OF *Clostridium botulinum* TYPE B (DAYS)

Temperature (°C)	*pH*	*Water activity*			
		0·997	0·98	0·97	0·96
20	6	1/4	1/9		
30	6	1/2	1/3	1/9	
40	6	1/1	1/2	1/3	
20	7	1/2	1/4	1/9	
30	7	1/1	1/2	1/3	1/9
40	7	1/1	1/1	1/2	1/3

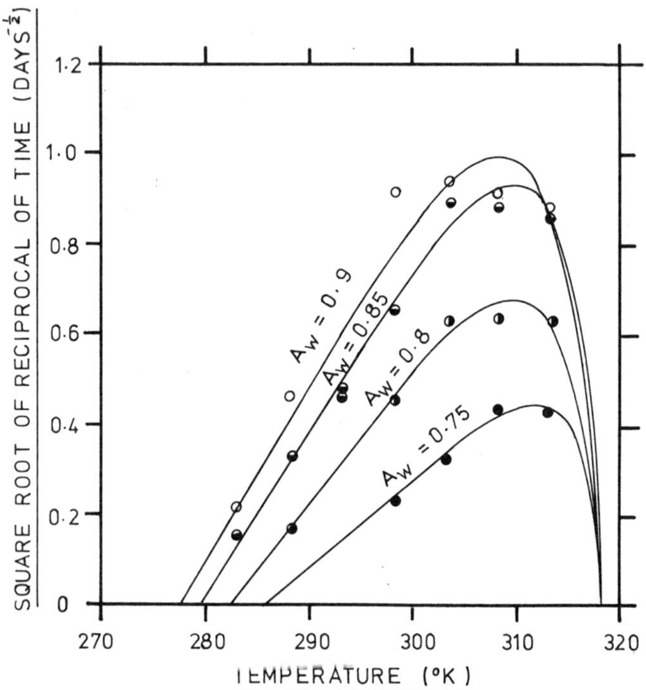

FIG. 4. Square root of the reciprocal of time for growth of colonies of *Aspergillus chevalieri* var. *intermedius* to grow to 1 mm diameter.

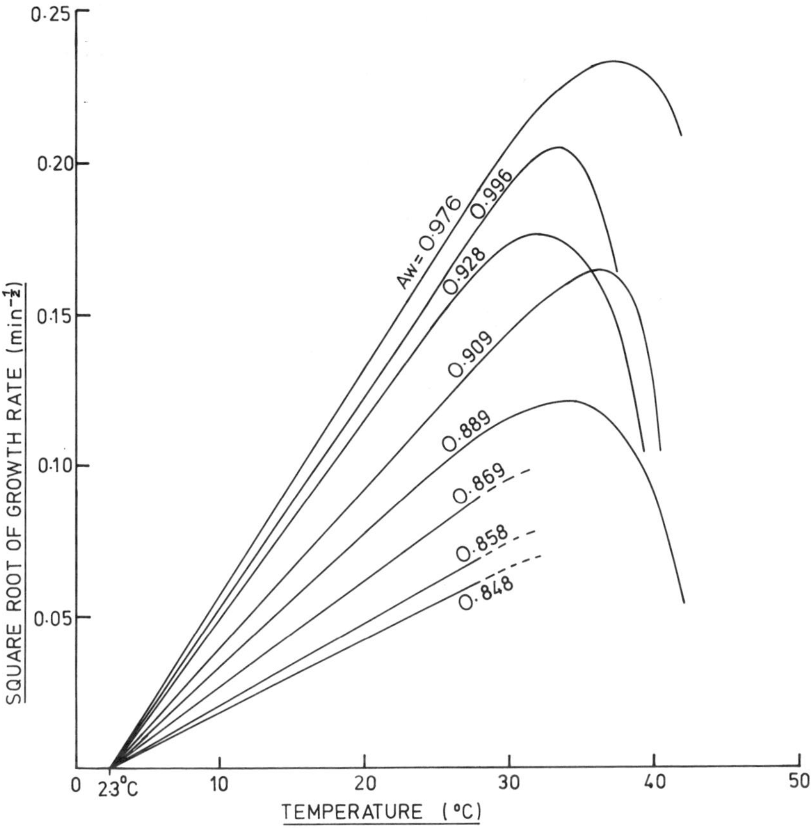

FIG. 5. Square root of reciprocal of germination time of *Staphylococcus xylosus* versus temperature at different water activities.

from 0·997 to 0·98. A similar effect at pH 7 was obtained by the same reduction in water activity but by reducing the temperature from 30°C to 20°C.

When the data of Table 1 are plotted as the square root of the reciprocal of the time to reach a visible colony, one obtains a series of straight lines up to the point at which thermal death becomes significant. Values of T_{min} for each water activity can be determined. A reduction in water activity and an unfavourable pH raise the temperature at which growth of the organism is totally prevented. Chilling is therefore more favourable under adverse conditions of water activity and pH.

TABLE 2

VALUES OF T_{min} AND b (EQN 1) FOR *Pseudomonas fluorescens* OVER A RANGE OF pH AND Aw (WODZINSKI AND FRAZIER, 1960)

T_{min}	*pH* 5·4 280·3 ± 1·5	*pH* 7·0 279·0 ± 1·0	*pH* 8·8 283·4 ± 1·1
Aw		*Values of b*	
0·9989			
0·9966	0·043 ± 0·004	0·033 ± 0·002	0·045 ± 0·004
0·990	0·033 ± 0·003	0·031 ± 0·002	0·036 ± 0·004
0·985	0·031 ± 0·003	0·026 ± 0·002	0·030 ± 0·003
0·980	0·022 ± 0·003	0·022 ± 0·002	0·022 ± 0·003
0·975	0·015 ± 0·002	0·018 ± 0·001	0·017 ± 0·003
0·970		0·012 ± 0·001	

Combined germination and growth data for the mould *Aspergillus chevalieri* from Ayerst (1969) are reproduced to extend the concept (Fig. 4). The rate of growth of this organism is halved by reducing the water activity from 0·9 to 0·82, or reducing the temperature from 30°C to 20°C. For this organism, reducing the water activity also appears to raise the value of T_{min} (Doe, 1983).

Some organisms appear to be able to withstand a large change in water activity and pH without a change in T_{min}. An extensive study of a spoiler of partially dried fish, *Staphylococcus xylosus*, indicated a constant T_{min} in the Aw range 0·996–0·846 (McMeekin *et al.*, 1987) (Fig. 5).

Pseudomonas fluorescens maintains its T_{min} within the range 6°C to 10°C over a range of pH from 5·4 to 8·8 and Aw from 0·97 to 0·9966 (Table 2).

The Time–Temperature–Water Activity Envelope

The concept of the combined effects of temperature and water activity on the growth rates of microorganisms being represented as a three-dimensional figure has been described by Doe (1985). When a number of different organisms are involved in fish spoilage, as would be the case in spoilage during drying, the three-dimensional time–temperature–water activity shapes for each organism overlap to form an envelope of all the shapes. If the fish is to be dried without spoiling, the temperature and water activities during drying must be such as to not intersect this spoilage envelope. An envelope based on organisms isolated from Indonesian dried fish is shown in Fig. 6.

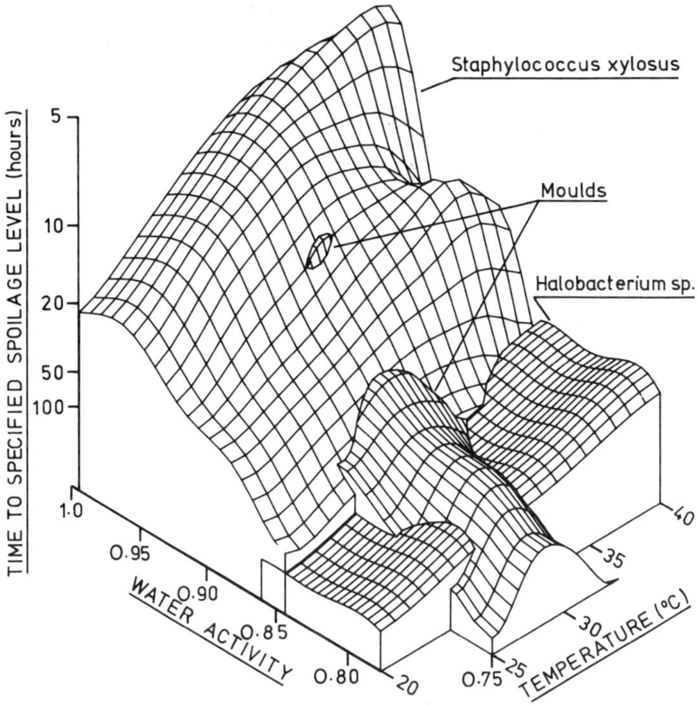

FIG. 6. Spoilage envelope based on measured growth rates of *S. xylosus*, *H. salinarium* and moulds isolated from Indonesian dried fish. Time to spoilage is on the basis of bacterial growth to 10^8 count and mould growth to 2 mm diameter colonies (Doe, unpublished data).

The envelope shows the ecological niche of the various bacteria and moulds. When the conditions of water activity and temperature favour one particular species it will predominate. An example is the spoilage of dried fish from Indonesia. At intermediate water activities (between Aw 0·95 and Aw 0·84) the bacterium *S. xylosus* which is usually a contaminant from human skin or domestic animals (Seager *et al.*, 1986) predominates; if the fish is dried to a water activity of 0·82 or less, however, a halophilic bacterium such as *Halobacterium salinarium* outgrows its competitors and characteristic pink spoilage occurs (R. E. Chandler and T. A. McMeekin, personal communication). The tables in Chapter 9 show the tremendous variability in Aw of the product with both species and region.

Deleterious Effects of Reducing Water Activity

The complex interactions of reducing sugars, or lipid oxidation products with proteins and free amino acids has been extensively covered in a previous IUNS/IUFoST workshop on Maillard reactions in foods (Eriksson, 1981). The browning of cod by Maillard type reactions was found to be at a maximum at a water activity of about 0·15 at 25°C and to have a broad maximum between Aw 0·15 to 0·6 at 60°C (Troller and Christian, 1978).

The role of Aw in the oxidation of lipids is controversial (Troller and Christian, 1978). Oxidation is maximised at very low water activities probably because of the concentration of metal catalysts. Labuza (1971) suggested that water interacts with metal catalysts to make them less effective and bonds to hydroperoxides preventing them from initiating chain reactions. The effect of water activity on lipid oxidation in foods has been extensively reviewed by Schaich (1980). At higher water activities still, the water present is sufficient to increase solution, diffusion and swelling of the solid matrix.

The reactions of oxidised lipids with amino acids and proteins leads to damage to the four most limiting amino acids in foodstuffs (Schaich, 1980). Cystine and methionine are damaged by the production of sulphoxides and sulphones and the indole ring of trytophane is damaged easily by oxidised fats. The formation of Schiff bases with the protein lysine has been extensively studied.

Rates of enzyme activity and oxidation are generally higher at water activities in the range 0·3–0·8. The discrepancies between the diagrammatic representations of Rockland and Nishi (1980) and Labuza (1971), however, show how few exact data are available for these reactions. Korean work on unsalted file fish has shown that non-enzymic browning is the result of lipid oxidation (Han *et al.*, 1983). Reaction rates at different water activities more closely resemble the Aw reaction rate diagram of Labuza (1971). Shelf life at 25°C assessed by accelerated shelf-life tests is 125, 123 and 106 days at Aw 0·44, 0·65 and 0·75 respectively (You and Lee, 1982).

Lowering the water activity of a fish product can lead to mould growth and the production of mycotoxins. Although there have been no reported incidences of consumers being poisoned by mycotoxins, the presence of mycotoxin in some fish products indicates that there is a risk to human health, particularly from some traditionally processed products (Reilly, 1986). *Aspergillus flavus* has been isolated from Indonesian dried fish, but was not growing on the fish and produced no aflatoxin (Anggawati *et al.*,

1984/85). It should be noted that not all moulds which occur on fish are considered deleterious; some fish curing practices are aimed at the production of specific moulds which, by outcompeting other possibly harmful moulds, improve the quality of the product (Tanikawa, 1985; see also Chapter 8).

Effects on Insects

The effect of temperature, water activity and salt on the attack of dried fish by flies, mites and beetles has been reviewed (FAO, 1981) and 'guesstimates' of losses are to be found in James (1983). The water activity requirements of larvae and adults are different and the contribution of the substrate is poorly understood. The viability of eggs, and the development of larvae both at 28°C and 30°C of *Dermestes maculatus*, a beetle commonly found in salted dried fish (FAO, 1981), are optimum in a range of moisture and salt contents from 11 to 15% and less than 4% respectively. Another commonly occurring beetle, *Necrobia rufipes*, prefers a lower salt content (Osuji, 1975, 1978) and the blowfly, species of *Chrysomya*, needs a higher moisture content.

The succession of infestation is directly related to the changing moisture content regimes in the fish during drying and subsequent storage. Initial infestation is usually by *Diptera* which are replaced by dermestids and clerid beetles during drying and storage. Tyroglyphid mites are also associated with dried fish at various stages during storage and marketing, especially during cool wet seasons (Osuji, 1977).

It has been suggested that a high lipid content may increase insect infestation; fat lowers water activity. Indonesian dried fatty fish (*Pangacius* sp.) were found to be infested mostly by *Chrysomya* and *Dermestes* (Indriati *et al.*, 1986). The Aw requirements are known for only a few insects (FAO, 1981), but it should be noted as with xerophilic moulds, that the maximum Aw may not be the preferred one. Osuji (1977) and Daniel and Etoh (1983) have recorded the progression of fly and beetle infestation in stored fish in Nigeria and Sri Lanka respectively. Water is a metabolite of flies, mites and beetles which may give rise to local increases in Aw, facilitating further microbial growth.

Proctor (1977) has claimed that flies are not attracted to salted fish and the degree of attraction of adult dermestid beetles is inversely related to the amount of initial salting; Esser *et al.* (1985) found, however, that *Arius* sp. brined in 25% salt were heavily infested by blowfly larvae during drying. Flies of *Musca* and *Chrysomya* spp. have a high possibility of *Enterobacteriaceae* transmission (Greenberg, 1971). Numerical considerations, however, cast doubt that *Dermestes* is involved significantly in the dissemination of *Salmonella* sp. (Julseth *et al.*, 1969).

Other Effects

Direct physical damage and loss can be caused by insect and animal predators. Fragmentation of hot smoked fish is a major source of loss. Birds, small animals and rodents can be a problem. Beetle infestation also causes an increase in fragmentation. The use of a solar drier is recommended (Sachithananthan *et al.*, 1986) as a means of reducing the losses due to birds etc., and also as a method of disinfesting fish from insects. Curran and Trim (1982) have successfully developed and tested solar fish driers of one tonne wet fish capacity.

CONTROL OF SPOILAGE

In simple terms, spoilage of fish can be retarded by subjecting it to conditions of temperature, water activity, pH, etc. which are not conducive to microbial growth, enzyme activity, oxidation or insect infestation. Salt is a good ally in many respects, but high levels of salt are considered undesirable today and are linked to hypertension (Wekell *et al.*, 1983; Shank *et al.*, 1983), a consideration which developing countries may have to ignore where food is of the first priority. Most cured fish, however, are desalted prior to consumption and where salt is left in, the fish is often used in small quantities to impart seasoning to certain local dishes.

Smoking will be dealt with in later chapters. It is the drying effect of the smoking process which lowers the water activity and hence contributes to the stability of the smoked product. Smoke is also an antioxidant. It is not considered to protect against insect infestation other than by drying the product (Proctor, 1977).

It may not, however, be economically viable or indeed possible to minimise all factors contributing to spoilage. Also, as mentioned above, there may be other influences, such as consumer preference, which dictate the method of preservation. An increase in the cost of the product due to increased processing costs may make the product unobtainable by the lower socio-economic groups.

No attempt has been made to discuss fungistats such as sorbic acid or insect repellants such as pyrethrum (Anon., 1977; Taylor, 1983; Esser *et al.*, 1985, and see Chapter 8). This chapter has been concerned with the fundamental parameters which can be manipulated during the transport of fish and its subsequent drying and storage.

REFERENCES

Aima, M.H., Fawzya, Y.N. and Putro, S. (1984). Effects of on board handling on the quality of dried salted *Clupea brachysoma* Blkr. *Res. Report Fish Technol. RIFT* **34**, 1–10.

Anggawati, A.M., Gorczyca, E., Hocking, A.D., Pitt, J.I., Tobin, N.F., Tuffs, L.C. and Wheeler, K.A. (1984/85). Spoilage in Indonesian dried fish. *Res. Report CSIRO Div. Food Res.* 33–4.

Anggawati, A.M., Gorczyca, E. and Putro, S. (1985). ACIAR Project 8304 – Prediction and control of spoilage of fresh, cured and dried fishery products in Indonesia: Commercial fish quality at retail markets in Jakarta. *Res. Report Fish Technol. RIFT* **44**, 5–11.

Anggawati, A.M., Hocking, A.D., Pitt, J.I. and Wheeler, K.A. (1985/86). Spoilage in Indonesian dried fish. *Res. Report CSIRO Div. Food Res.* 3–4.

Anonymous (1977). Discussion, Session 6. Canned and other processed fish. In: *Handling, Processing and Marketing of Tropical Fish*, P. Sutcliffe and J. Disney (Eds). Tropical Products Institute, Ministry of Overseas Development, London, p. 357.

Ayerst, G. (1969). The effects of moisture and temperature on growth and spore germination in some fungi. *J. Stored Prod. Res.* **5**, 127–41.

Belehrádek, J. (1935). In: *Temperature and Living Matter*, Protoplasma Monographs No. 8, Borntraeger, Berlin, 277 pp.

Branch, A.C. and Vail, A.M.A. (1985). Bringing fish inspection into the computer age. *Food Technol. Aus.* **37**(8), 352–5.

Bremner, H.A., Kearney, P., Olley, J., Quarmby, A.R., Statham, J.A., Sykes, S.J. and Vail, A.M.A. (1984/85). Shelf life of species from the North-West shelf: Finfish. *Res. Report CSIRO Div. Food Res.* 58–9.

Chou, H. and Labuza, T.P. (1974). Antioxidant effectiveness in intermediate moisture content model systems. *J. Food Sci.* **39**, 479–83.

Clucas, I.J. and Sutcliffe, P.J. (1981). *An Introduction to Fish Handling and Processing*, Tropical Products Institute, London.

Cruickshank, E.M. (1962). Fat soluble vitamins. In: *Fish as Food* G. Borgstrom (Ed.), Academic Press, London, Vol. 2, p. 175.

Curran, C.A. (1984). *Storage of cured fishery products in Malaysia*, Tropical Development and Research Institute, London, 67 pp.

Curran, C.A. and Poulter, R.G. (1983). Isohalic sorption isotherms III. Application to a dried salted tropical fish (*Xenomogil thoburni*). *J. Food Technol.* **18**, 739–46.

Curran, C.A. and Trim, D.S. (1982). Comparative study of three solar fish driers. In: *Proceedings of the FAO Expert Consultation on Fish Technology in Africa*, FAO Fisheries Report 268 Suppl., FAO, Rome, 288 pp.

Daniel, D.J.E. and Etoh, S. (1983). Insect infestation on dried fish in Sri Lanka.

In: *The Production and Storage of Dried Tropical Fish*, D. James (Ed.), FAO Fisheries Report 279 Suppl. FAO, Rome, pp. 162–71.
Doe, P.E. (1983). Spoilage of dried fish – the need for more data on water activity and temperature effects on spoilage organisms. In: *The Production and Storage of Dried Tropical Fish*, D. James (Ed.), FAO Fisheries Report 279 Suppl. FAO, Rome, pp. 209–15.
Doe, P.E. (1985). An integrated approach to fish spoilage. In: *Proceedings of Food Conference 1985*, E.F. Alabastro, T.P. Acevedo and L.L. Chavez (Eds), ASEAN Food Research Program, Philippines, 527 pp.
Doe, P.E., Hashmi, R., Poulter, R.G. and Olley, J. (1982). Isohalic sorption isotherms I. Determination for dried salted cod (*Gadus morrhua*). *J. Food Technol.* **17**, 125–34.
Doe, P.E., Curran, C.A. and Poulter, R.G. (1983). Determination of the water activity and shelf life of dried fish products. In: *The Production and Storage of Dried Tropical Fish*, D. James (Ed.), Fisheries Report 279 Suppl. FAO, Rome, pp. 202–8.
Dyerberg, J. and Jørgenson, K.A. (1982). Marine oils and thrombogenesis. *Prog. Lipid Res.* **21**, 255–69.
Eriksson, C. (Ed.) (1981). Maillard reactions in foods. In: *Progress in Food and Nutrition Science*, Pergamon Press, New York, Vol. 5, Nos 1–6, 501 pp.
Esser, J.R., Hanson, S.W. and Taylor, K.D.A. (1985). Loss reduction techniques for salted/dried marine catfish (*Arius* sp.). In: *Spoilage of Tropical Fish and Product Development*, A. Reilly (Ed.), FAO Fisheries Report 317 Suppl. FAO, Rome, 474 pp.
FAO (1981). The prevention of losses in cured fish. Fisheries Technical Paper No 219, FAO, Rome, 87 pp.
Gildberg, A. (1982). Autolysis of fish tissue – general aspects. Doctor of Science degree. Institute of Fisheries, University of Tromsø, Norway.
Greenberg, B. (1971). In: *Flies and Disease. Ecology, Classification and Biotic Associations*, Princeton University Press, Princeton, New Jersey, Vol. 1, 389 pp.
Han, B.H., Lee, J.G. and Bae, T.J. (1983). Studies on food preservation by controlling water activity III. Quality changes of fish meat during drying and storage. *Bull. Korean Fisheries Soc.* **16**(3), 181–9. (*Food Science and Technology Abstracts* 1985(5) p. 191.)
Hoffman, A. (1981). The use of the GR Torrymeter for the assessment of freshness of iced tropical fish from the Indian Ocean. *Tropical Sci.* **23**, 283–91.
Indriati, N.Y., Sudrajat, Y., Anggawati, A.M., Madden, J. and Putro, S. (1986). Insect infestation in dried salted fish in Java. *J. Postharvest Fisheries Res.* **50**, 9–12.
James, D. (Ed.) (1983). The production and storage of dried fish. In: *Proceedings of the Workshop on the Production and Storage of Dried Fish*, Universiti Pertanian, Serdang, Malaysia, 2–5 November 1982, FAO Fisheries Report 279 Suppl. FAO, Rome, 265 pp.
James, D. (1984). The future for fish in nutrition. *Infofish Marketing Digest* **4**, 41–4.
Julseth, R.M., Felix, J.K., Burkholder, W.E. and Deibel, R.H. (1969). Experi-

mental transmission of *Enterobacteriaceae* by insects. *Appl. Microbiol.* **17**, 710–13.
Karel, M. (1984). Chemical effects in food stored at room temperature. *J. Chem. Educ.* **61**(4), 335–9.
Labuza, T.P. (1971). Kinetics of lipid oxidation in foods. *CRC Crit. Rev. Food Technol.* **2**, 355–405.
Labuza, T.P. (1984). Application of chemical kinetics to deterioration of foods. *J. Chem. Educ.* **61**(4), 348–58.
Love, R.M. (1970). Depletion. In: *The Chemical Biology of Fishes*, Academic Press, London, pp. 222–57.
Love, R.M. (1982). Basic facts about fish. In: *Fish Handling and Processing*, A. Aitken, I.M. Mackie, J.H. Merritt and M.L. Windsor (Eds), HMSO, Edinburgh, p. 2.
Lupin, H.M. (1986). Water activity in preserved fish products. In: *Cured Fish Production in the Tropics*, A. Reilly and L.E. Barile (Eds), College of Fisheries, University of the Philippines, pp. 16–55.
McMeekin, T.A., Chandler, R.E., Doe, P.E., Garland, C.D., Olley, J., Putro, S. and Ratkowsky, D.A. (1987). Model for combined effect of temperature, salt concentration/water activity on the growth rate of *Staphylococcus xylosus*. *J. Appl. Bacteriol.* **62**, 543–50.
Moschair, S.M., Fardin, J.P. and Boeri, R.L. (1984). Sorption isotherms of dried salted hake (*Merluccius hubbsi*). *Lebensm.-Wiss. u.-Technol.* **17**, 86–9.
Nettleton, Joyce A. (1985). In: *Seafood Nutrition. Fact, Issues and Marketing of Nutrition in Fish and Shellfish*, Osprey Books, Huntington, New York, 280 pp.
Ohta, F. and Hirahara, T. (1977). Rate of degradation of nucleotides in cool-stored carp muscle. *Mem. Fac. Fisheries, Kagoshima Univ.* **26**, 97–102.
Olley, J. (1983). Temperature effects on histamine producing bacteria. In: *Properties and Processing of Marine Foods*, Bonnie Sun Pan (Ed.), National College of Marine Science and Technology, Taiwan, pp. 14–35.
Olley, J. and Lisac, H. (1985). Time-temperature monitors. *Infofish Marketing Digest* **3**, 45–7.
Olley, J. and Ratkowsky, D.A. (1973). Temperature function integration and its importance in the storage and distribution of flesh foods above the freezing point. *Food Technol. Aus.* **25**(2), 66–73.
Osuji, F.N.C. (1975). The effects of salt treatment on fish on the developmental biology of *Dermestes maculatus* Deeger (Coleoptera, Dermestidae) and *Necrobia rufipes* (Cleridae). *Entomol. Exp. Appl.* **18**, 472–9.
Osuji, F.N.C. (1977). The influence of traditional handling methods on the quality of processed fish in Nigeria. In: *Handling, Processing and Marketing of Tropical Fish*, P. Sutcliffe and J. Disney (Eds.), Tropical Products Institute, Ministry of Overseas Development, London, pp. 307–11.
Osuji, F.N.C. (1978). An assessment of the performance of *Dermestes maculatus* Deeger (Coleoptera, Dermestidae) in some dietary media. *Entomol. Exp. Appl.* **24**, 185–92.
Owen, D. and Nesbitt, M.A. (1984). A versatile time-temperature function integrator. *Lab. Pract.* **33**, 30–75.

Pitt, J.I. and Hocking, A.D. (Eds) (1985). *Fungi and Food Spoilage*, Academic Press, Sydney, 413 pp.

Pooni, G.S. and Mead, G.C. (1984). Prospective use of temperature function integration for predicting the shelf life of non-frozen poultry-meat products. *Food Microbiol.* **1**, 67–78.

Proctor, D.L. (1977). The control of insect infestation in fish during processing and storage in the tropics. In: *Handling, Processing and Marketing of Tropical Fish*, P. Sutcliffe and J. Disney (Eds), Tropical Products Institute, Ministry of Overseas Development, London. pp. 307–11.

Ratkowsky, D.A., Olley, J., McMeekin, T.A. and Ball, A. (1982). Relationship between temperature and growth rate of bacterial cultures. *J. Bacteriol.* **149**(1), 1–5.

Ratkowsky, D.A., Lowry, R.K., McMeekin, T.A., Stokes, A.N. and Chandler, R.E. (1983). Model for bacterial growth throughout the entire biokinetic temperature range. *J. Bacteriol.* **154**, 1222–6.

Reilly, A. (1986). Mycotoxins in seafood. In: *Cured Fish Production in the Tropics*, A. Reilly and L.E. Barile (Eds), College of Fisheries, University of the Philippines, pp. 131–8.

Rockland, L.B. and Nishi, S.K. (1980). Influence of water activity on food product quality and stability. *Food Technol.* **34**(4), 42–51, 59.

Ross, K.D. (1975). Estimation of water activity in intermediate moisture content foods. *Food Technol.* **29**(3), 26–34.

Sachithananthan, K., Trim, D.S. and Speirs, C.I. (1986). A solar-dome dryer for drying fish. In: *Fish Processing in Africa*. FAO Fisheries Report 329 Suppl. FAO, Rome, pp. 161–72.

Schaich, H.M. (1980). Free radical initiation in proteins and amino acids by ionising radiations and lipid oxidation – Part III, Free radical transfer from oxidising lipids. *CRC Crit. Rev. Food Sci. Nutr.* **13**, 189–244.

Seager, M.S., Banks, J.G., Blackburn, C. de W. and Board, R.G. (1986). A taxonomic study of *Staphylococcus* spp. isolated from fermented sausages. *J. Food Sci.* **51**(2), 295–7.

Shank, F.R., Larsen, L., Scarbrough, F.E., Vanderveen, J.E. and Forbes, A.L. (1983). FDA Perspective on sodium. *Food Technol.* **37**(7), 73–5, 77.

Sperber, W.H. (1983). Influence of water activity on foodborne bacteria – a review. *J. Food Prot.* **46**(2), 142–50.

Sumner, J.L., Gorczyca, E., Cohen, D. and Brady, P. (1984). Do tropical fish spoil less rapidly in ice than temperate fish? *Food Technol. Aus.* **36**(7), 328–9, 334.

Tanikawa, E. (1985). *Marine Products in Japan* (revised edition), Koseisha-Koseikaku Company, Tokyo, p. 250.

Taylor, R.W.D. (1983). Insecticides for the protection of dried fish. In: *Proceedings of the FAO Expert Consultation on Fish Technology in Africa*, Casablanca, Morocco, 7–11 June 1982. FAO Fisheries Report 268 Suppl. FAO, Rome, p. 357.

Troller, J.A. and Christian, J.H.B. (Eds) (1978). *Water Activity and Food*, Academic Press, New York, 235 pp.

Wekell, J.C., Teeney, F.M., Gauglitz Jr., E.J., Hathorn, L. and Spinelli, J. (1983).

Implications of reduced sodium usage and problems in fish and shellfish. *Food Technol.* **37**(9), 51–8, 63.

Wodzinski, R.J. and Frazier, W.C. (1960). Moisture requirements of bacteria. 1. Influence of temperature and pH on requirements of *Pseudomonas fluorescens*. *J. Bacteriol.* **77**, 572–8.

You, B.J. and Lee, K.H. (1982). Kinetics of lipid oxidation in dried fish flesh stored under different conditions of water activity and temperature. *Bull. Korean Fisheries Soc.* **15**(1), 83–93. (*Food Science and Technology Abstracts* 1983(12), 225–6.)

Chapter 2

INFLUENCE OF DRYING AND SMOKING ON PROTEIN QUALITY

J. OPSTVEDT
Norwegian Herring Oil and Meal Industry Research Institute, Norway

INTRODUCTION

Methods of drying and smoking of fish vary between different countries and within the same country depending on the species of fish used and the type of product desired. The process may use unsalted fish or various salt additions giving salt concentrations in the final products ranging from less than 2% to over 20%. In addition other curing agents (e.g. nitrate) may be added. The fish may be dehydrated to various degrees with moisture levels in the final product ranging from about 10% to 60%. Processing temperatures may range from less than 5 to up to 120°C and times from half an hour to several months. The fish may be dried only or smoked only or there may be a combination of smoking and drying. In some countries the fish is boiled before being smoked and/or dried. Adding to this complexity the fish species used as raw material may be freshwater or marine species and may range from very lean to fatty fishes, and its condition from fresh to stale. This variation makes it difficult to arrive at general conclusions regarding processing effects of smoking and drying on protein quality and the nutritional value of the final products. Evidently, the impact of drying and smoking on protein quality may be assessed from known effects of physical/chemical parameters employed in the production such as dehydration, temperature, time, added compounds and oxidation, but only tests on the actual products can provide reliable data on their protein quality.

The effect of processing on protein quality has previously been discussed by Bender (1972, 1977), Carpenter (1973), Cutting (1962) and Lionarons (1970). This paper attempts to gather and evaluate data from experiments conducted in relation to practical production and with emphasis on the situation in developing countries.

Protein quality can be measured or assessed in a variety of different ways using *in vitro* or *in vivo* tests. Those tests referred to in this chapter are described and defined at the end of it.

EFFECTS ON PROTEIN QUALITY

Chemical Reactions in Proteins

The heating of proteins causes denaturation which is a rupture of the secondary and higher structures (Ledward, 1979). The temperature at which the denaturation occurs (i.e. denaturation temperature) varies for the different fish proteins and for the same protein between different fish species, being related to the environmental temperatures in which the fish live (Aitken and Connell, 1979). Generally speaking about 90% of the protein is denatured at about 60 to 65°C, the remaining 10% (tropomyosin) may be held at 100°C for a prolonged period of time without being denaturated. Denaturation may also occur over time at lower temperatures i.e. during frozen storage. Denaturation *per se* does not appear to affect protein utilisation by animals in a negative manner. Thus, freeze-drying of fish which causes extensive denaturation of the actomyosin (Connell, 1962), does not reduce protein quality compared with that of the raw native fish protein (Adachi *et al.*, 1958; Thomas and Calloway, 1961; Yasui, 1974; Maekawa *et al.*, 1975; and Opstvedt *et al.*, 1984).

As a consequence of denaturation, proteins may aggregate; this involves the formation of new bonds, some of which are assumed to be covalent (Ledward, 1979). Thus, Opstvedt *et al.* (1984) found a linear decrease in the content of —SH (sulphydryl) groups and a concomitant increase in the content of S—S (disulphide) bonds when fish is heated at increasing temperatures from 50 to 115°C. Formation of secondary intramolecular disulphide bonds may also occur over time during frozen storage (Buttkus, 1970). The impact of disulphide bond formation on protein utilisation is not fully known, but limited experimental data indicate that it may reduce protein digestibility (Opstvedt *et al.*, 1984).

Reactions of significance for protein utilisation may occur between the ϵ-NH_2 groups of lysine and reducing non-protein compounds through carbonyl-amine condensation. These reactions may occur at ambient or lower temperatures over time, although the rate at which they occur increases with temperature. The reaction is best described with reducing sugars, but these substances are present in fish at levels too low to be of practical significance (Carpenter *et al.*, 1963; Nagayama, 1960). Carbonyl compounds present in smoke (Ziemba, 1967; Obileye and Spinelli, 1978) or arising from lipid oxidation (Gardner, 1979), however, may react with the ϵ-NH_2 of lysine in a similar manner to reducing sugars (see also Baba, 1960). Further, sodium nitrate, commonly used as a curing substance together with salt, may react with lysine (Wilder and Kraybill, 1947). The impact of these reactions is a reduction in protein utilisation, in particular in lysine availability, by animals, and is discussed in detail by Carpenter (1973).

At higher temperatures protein suffers thermal degradation. Thus at 115°C losses of cysteine/cystine occur both at low (14%) (Bjarnarson and Carpenter, 1970) and high (70–80%) (Opstvedt *et al.*, 1984) moisture content. Thus, some 10% of the cystine/cysteine fraction is lost in fish heated in a retort at 115°C for 20 min (Opstvedt *et al.*, 1984).

Heating of proteins at temperatures above 115°C, i.e. up to 145°C, causes extensive destruction of several amino acids (Ledward, 1979). The ϵ-NH_2 group of lysine may react with asparagine and glutamine and with breakdown products of cystine to form secondary intramolecular covalent bonds (Bjarnarson and Carpenter, 1970). The ultimate result of these reactions is severe depression of protein utilisation (Shorrock and Ford, 1978).

Influence of Cooking

Cooking is sometimes used as a pre-treatment in the processing of dried-smoked fish products and some dried-smoked fish products are cooked immediately before they are eaten. Consequently, it is pertinent to assess the effect of cooking of fish on protein quality.

Baba (1960) found that heating hen clam at 85°C causes a reduction in the content of FDNB (1-fluoro-2,4-dinitro-benzene)-reactive lysine of 3·5% after 2 h and of 13% after 96 h.

Sawant and Magar (1961) tested the effect of canning (15 lb/in^2: 100°C) on the content of amino acids in white pomfret and prawns. Cleaned pomfret was cut into pieces and prawns had their heads and shells removed and were blanched in hot water. They were then canned in 2·5% brine. Losses in total amino acids range from 12 to 0%, being greatest for lysine in pomfret and least for phenylalanine, iso-leucine and arginine in prawns. The differences between pomfret and prawns are not consistent and the contents of cystine were not analysed.

Nagayama *et al.* (1962) observed that the autoclaving at 110°C of sole, sea bass and cod, freshly produced or stored for 24 and 72 h before being processed, reduces the content of arginine by 3·5 to 18·1% and the content of lysine by 6·1 to 17·1 %. Autoclaving at 120°C reduces arginine by 3·9 to 41·0 and lysine by 10·4 to 24·0 %.

Yabuki and Izumi (1966) found that glycine, alanine, leucine, glutamic acid, valine and serine are present in the cooking water of both fish and shellfish, while aspartic acid and glutamine are found in the cooking water of shellfish only. The quantity of these amino acids lost in the cooking water is not given.

Krauze *et al.* (1970) reported that the *in vitro* digestibility of different fish species decreased in the following order: fish fried in oil, boiled fish and raw fish.

Kennedy and Ley (1971) tested the effects of cooking cod fillets for 4 min at 15 lb/in^2 (about 100°C) in a pressure cooker. Compared with raw fillets cooking reduces protein quality by 9% when assayed by a microbiological test (RNV = relative nutritive value).

The nutritive effects of cooking frozen stored tuna meat were studied by Seet and Brown (1983) using various *in vitro* assays. Compared with raw flesh, cooking at 100°C for 3 h reduces the content of lysine, histidine and cystine by 3 to 4%, but has no effect on the other amino acids. Also, cooking reduces *in vitro* protein digestibility by 0·8 %, and FDNB-reactive lysine by 9%, but has no effect on the computed PER (protein efficiency ratio) figure or in the *Tetrahymena* test. Autoclaving at 115° and 121°C exaggerates the effects found at 100°C.

Opstvedt *et al.* (1984) cooked minced pollock and mackerel for 20 min after the internal temperature had reached 95°C. Compared with the raw fish, cooking reduces the digestibility of protein in rainbow trout by 3% and of amino acids by some 2% in pollock, but has no consistent effects in mackerel.

Opstvedt *et al.* (1987) observed that heating of minced fish in water for 20 min at 55°C did not affect protein digestibility, while heating at 95 °C reduced protein digestibility in pre-ruminant calves by 6·7% compared with that of raw fish. Heating had no effect on the utilisation of the absorbed protein.

The results from Cesare (1980) are in contrast to these observations. The latter author heated sardines to 50°, 100°, 150° and 200°C for increasing periods of time from 7·5 to 240 min in a factorial arrangement. Compared with the unheated control, heating for 7·5 min increased the content of FDNB-reactive lysine at all temperatures although slightly less was produced at 200° than at 150°C. The content, on a dry matter basis, of FDNB-reactive lysine increased with heating time for up to 240 min at 50°C and for up to 15 min at 100°C with decreases thereafter. Heating times in excess of 7·5 min reduced the content of FDNB-reactive lysine at both 150° and 200°C. One may wonder whether losses of volatile reactive substances during the heating may have affected the results.

Although the experimental results are not uniform, it appears that normal cooking procedures using temperatures up to 100°C have only a slightly detrimental effect on protein quality, and that this negative effect is more pronounced when pressure cooking at temperatures above 100°C is used. The effect may partly be due to losses of amino acids in the cooking water, but in addition there may be a moderate (about 5%) reduction in amino acid availability (digestibility).

Influence of Drying Unsmoked Fish

Proctor and Lahiry (1956) tested the effect of dehydration on the content of amino acids in shad and haddock. Fillets were cooked for 10 min at 4 lb/in^2 (100°C) and thereafter dried in a vacuum belt dryer for 300 min at 88°C followed by 250 min at 77°C. There are no consistent, significant differences in the content of amino acids between raw and dehydrated fish, but valine, phenylalanine and tryptophan tend to be somewhat lower in the dehydrated compared with the raw fish. Cysteine was not analysed.

Adachi *et al.* (1958) freeze-dried (plate temperature 38°C) haddock fish sticks over 28–29 h. Freeze-drying does not affect the PDR (pepsin digest residue) amino acid index either in the raw or in the pre-cooked form. Dried fish retains its PDR-index after storage at temperatures up to 38°C for up to 12 months.

Thomas and Calloway (1961) studied the effect of dehydration to 8–10% moisture content by freeze-drying raw and pre-cooked shrimp using various nutritional criteria. Freeze-drying does not show any consistent effects on the content of amino acids either in raw or pre-cooked shrimp. A slight reduction in the amount liberated by *in vitro* pepsin digestion in dehydrated compared with wet samples is seen for most amino acids. Since only a small fraction (6% for lysine) is also liberated in the wet sample, the significance of this finding is unclear.

De Groot (1963) compared the protein quality in rats of undried and freeze-dried (plate temperature 46°C for 18 h cooked) (100°C for 1 h 35 min) fish. True digestibility (TD), biological value (BV) and net protein utilisation (NPU) levels found for undried and freeze-dried samples were 83·0 and 84·3, 100·0 and 101·1, and 83·1 and 85·3 respectively.

Mustafa (1966) studied the effect on protein quality of temperature and time in the drying of (homogenised) cod fillet. The TD values in rats and FDNB-reactive lysine concentration were found to be for freeze-dried

(21 h) samples 99·7 and 8·0, for vacuum-dried (40°C for 19 h) 98·0 and 8·1, for low-temperature oven-dried (70°C for 18 h) 98·0 and 7·9, and for high-temperature oven-dried (100°C for 16 h) samples 95·5 % and 7·0 g/16 g N respectively.

The effects of drying temperature and salting on the protein quality of dried cod were investigated by Aitken *et al.* (1967), using the AMD (accelerated mechanical drying) system. Unsalted fish have NPU_{st} (net protein utilisation standardised) values of 96 when dried with a maximum temperature of 100°C compared with 87 when dried with a maximum temperature of 115°C. Lightly (10·8 % NaCl) salted and heavily (28·9 % NaCl) salted fish have NPU_{st} values of 94 and 90 respectively after freeze-drying, compared with 90 for undried, lightly salted fish. The individual variations in the test results limit the possible conclusions.

Rats fed unsalted, air-dried coalfish (stockfish) had apparent protein digestibility and nitrogen balance of 85·2 and 59·5 % respectively in an experiment by Njaa *et al.* (1968). These values compare favourably with those for rats fed raw cod fillets and having apparent protein digestibility and nitrogen balance of 86·0 and 47·9 % respectively.

Konosu and Matsuura (1962) claimed that the tryptophan content in salt-dried fish meat is almost the same as in raw fish. This is in contrast to the results of Shinno *et al.* (1965) who found heating and drying reduced the total content of tryptophan, although they did not affect the *in vitro* protein digestibility by pepsin.

Yanes *et al.* (1970) dried hake fillets by freeze-drying or by oven-drying at 105 or 170°C. The figures for true protein digestibility are 94·3, 89·3 and 74·2; for biological value they are 82, 88 and 68; for net protein utilisation 79, 77 and 51; for true digestibility for sulphur 90, 82 and 66; for sulphur retained (g per 100 g N consumed, in rats) 1·0, 0·9 and 0·7; and for the FDNB-reactive lysine (g per 16 g N) 8·6, 8·3 and 6·9 respectively for the three types of dried product.

Herborg *et al.* (1974) tested the protein quality of roller-dried raw cod mince. The drying time was 25 s with a roller temperature of 155°C giving a final temperature of the fish at the end of the drying process of about 80°C. Roller-dried fish have true digestibility and biological value in rats of 97·8 and 84·6 compared with 96·4 and 77·7 respectively for the raw flesh.

Sheikh and Shah (1974) studied the effect of drying temperature on the protein quality of a fatty fish (Mahle) and relatively lean fish (Rahu). The homogenised mince was placed in a forced draught oven in 1 cm thick layers at 60, 80 or 100°C or was roller-dried. *In vitro* pepsin, trypsin and

TABLE 1

DIFFERENCES IN THE *in vitro* DIGESTIBILITY AND TRYPSIN INDIGESTIBLE SUBSTRATE (TIS) CONTENT OF RAW SEAFOODS AS A FUNCTION OF THE DRYING PROCESS

	Raw		*Sun-dried*		*Hot air-dried*		*Freeze-dried*	
	In vitro dig. (%)	*TIS (mg/g)*	*In vitro dig.* (%)	*TIS (mg/g)*	*In vitro dig.* (%)	*TIS (mg/g)*	*In vitro dig.* (%)	*TIS (mg/g)*
Californian market squid	84·4	10·8	83·0	12·8	81·4	11·4	84·7	12·7
Atlantic pollock	85·8	17·8	84·7	29·0	84·7	21·2	86·2	17·8
Flounder	87·6	13·9	85·5	15·5	86·6	14·7		
Hairtail fish	86·1	12·0	85·2	16·0	84·8	17·3		
Anchovy	78·5	8·9	84·6[a]	10·1[a]	84·2[a]	10·3[a]	85·9[a]	12·1[a]
Yellow corvina	84·7	14·0			88·0[b]	10·2[b]		
							86·7[c]	10·3[c]
Oyster	80·2	9·5	74·8	33·2	78·7	20·4	82·0	10·2
Salad shrimp[d]			84·8	15·6	86·8	17·0	89·3	9·3

[a] All dried anchovy products were made from boiled anchovy.
[b] 10% brine salting.
[c] 10% dry salting.
[d] Commercial boiled product.
After Lee and Ryu (1987).

pancreatic extract digestibility are highest in raw fish and decrease with drying temperature, roller-dried fish having digestibilities close to those of 80°C oven-dried fish. Consistent differences were not found between the fatty and the lean fish.

El-Wakeil *et al.* (1975) studied the protein quality of keshr bayad freshwater fish (*Lates niloticus*). In addition to salt (30% NaCl in DM), boric acid and sodium nitrate were added to the samples. Drying of the salted fish was done in air over 10 to 30 days depending on fish size, thereafter the fish was stored at room temperature for 30 days before being tested. The final product contained 10 to 12% moisture, about 12% of fat and about 80% protein (salt excluded). The protein quality of the salt-dried fish was compared with that of fresh fish after both samples had been dried at 105°C. NPU values for fresh fish, salt-dried fish and casein are 73·2, 69·5 and 61·7 respectively.

The effect of drying regimes on the *in vitro* digestibility of a number of seafood products has been tested by Lee and Ryu (1987) (see Table 1). Drying in the sun for a long time causes a slow, but only slight, lowering of digestibility due to protein damage and to the formation of enzyme indigestible substrates. Drying in a hot air blast (55°C) also causes a decrease in the digestibility of 'raw' seafoods but an increase in that of pre-cooked or salted products; boiling and salting appear to extract lipids and reduce the formation of enzyme-indigestible material. Freeze-drying causes smaller changes in digestibility than do other methods of drying and freeze-dried products appear to be slightly more digestible, even though they contain more trypsin-insoluble substrate (TIS), than the corresponding raw material.

From these results and others, drying (dehydration) *per se* may not have any adverse effect on protein quality since freeze-drying or vacuum-drying at low (below 60°C) temperatures has not been shown to reduce either content or availability of amino acids. Furthermore, as long as the drying temperature is kept below 100°C, losses in protein quality are moderate. When drying temperatures are in excess of 100°C, losses in protein quality increase progressively with temperature. It thus appears that the adverse effects of processing temperature are the same in cooking and drying, which indicates that the moisture content of the fish is of minor importance in this respect.

Influence of Smoking

Katsuobushi is a dried product made from bonito meat by a boiling, smoking, sun-drying and moulding process (see Chapter 8). According to Konuso and Hashimoto (1959) there is a reduction of 40% in the content of α-amino-N in DM of the final product compared with fresh fish, most of which occurs during the boiling step.

Munro and Morrison (1965) tested the protein quality of medium (48 % moisture and 20 % NaCl) salted and cold smoked commercial samples of cod in comparison with fresh cod. All samples were cooked and freeze-dried before assay with rats and humans. Differences were not found between salted and smoked cod and fresh cod with regard to content of total and FDNB-reactive lysine, or protein quality in rats (protein efficiency ratio and gross protein value) or humans (fasting plasma lysine and methionine levels). These results confirmed and extended previous studies on cold-smoked herring by Yoshikawa and Tamamoto (1942) in which protein digestibilities were found to be 96·4 and 97·6 %, in the cold-smoked and fresh samples respectively.

Ziemba (1967) concluded from his work that smoke carbonyls react with amino acids and that oxidative browning may take place if the smoke is produced with limited amounts of air.

Tahiri-Zagret (1969) analysed the content of amino acids and tested the protein quality of five samples of traditionally smoked fish (*Sphyraena*, *Trichiurus lepturus* and *Ethmalosa dorsalis*), in feeding tests with rats, in the Ivory Coast. For comparison a sample of commercial herring meal was included. All samples of smoked fish had an excellent protein quality which was equal to or better than that of the commercial herring meal.

The comprehensive work of Hoffman *et al.* (1977) is of considerable interest since it combined a detailed and controlled study of processing factors involved in smoking with an assay of commercially produced smoked fish in Africa. Two types of unsalted freshwater fish (*Tilapia* spp., about 7 % ether-extractable fat in DM) were smoked under different conditions (i.e. 65 h at a maximum temperature of 98°C, and 38 h at 50, 75 or 100°C). For comparison a sample was dried at 75°C. Commercial samples were smoked for 35 h at 100–112°C, or 4 h 10 min at 95–111°C and consisted of freshwater fish in addition to one sample of sardine (*Sardinella aurita*, 28 % ether-extractable fat in DM).

The different experimental treatments have no consistent effect on the contents of amino acids which also are similar in one of the commercial samples tested. Compared with the control, freeze-dried sample all smoking procedures reduce the content of FDNB-reactive lysine; this was also confirmed by a reduced content of FDNB-reactive lysine in two of the commercial samples tested. The magnitude of the reduction is related to the smoking conditions, i.e. about 25 % for 65 h at 98°C, 7, 8 and 9 % for 38 h at 50, 75 and 100°C. Storage, up to 3 months, of the smoked product does not cause additional losses in FDNB-reactive lysine. Drying at 75°C has somewhat greater effects on FDNB-reactive lysine than does smoking at a similar temperature. The effect of the treatments on the NPU values in rats are inconsistent. Two commercial samples were found to have NPU values of 49 and 69.

Although working with minced flesh of a freshwater fish (*Tilapia aurens*), the study of Obileye and Spinelli (1978) gives some indication of the effects of traditional curing of fish by salting, drying and smoking. The minced flesh was partly solubilised by the addition of 2% of salt plus tripolyphosphate and potassium sorbate, and moulded into sheets before being smoked for 2 h at 30°C and for an additional 2 h at 70°C to a moisture content of about 50% or for 12 h to a moisture content of about 20%. Smoking reduces the content of FDNB-reactive lysine by about 18% with little difference between different lengths of smoking time and ultimate moisture content. The authors noted that heating with the addition of liquid smoke causes greater loss in FDNB-reactive lysine than heating alone and concluded that reactive substances in the natural smoke react with the epsilon group of lysine.

Umoh *et al.* (1980) found locally produced, commercial smoked freshwater fish (*Tilapia melanopleura*) and dried crawfish (*Palamonetes varians*) to have *in vitro* pepsin–trypsin digestibilities of 75·3 and 69·9% compared with 96·9% for whole hen's egg. TD and BV in rats for smoked freshwater fish, crawfish and whole hen's egg are 91·1 and 70·8, 89·5 and 81·6, and 98·2 and 91·8, respectively.

Despite experimental data being scarce it appears that the presence of aldehydes in the smoke may reduce protein quality and in particular lysine availability. When the smoking process involves the application of high temperatures (above 100°C) additional adverse heat effects will be present.

CONCLUSIONS

Experimental data regarding the effect on nutritional protein utilisation of drying and smoking of fish and the quality of the products made thereby are extremely meagre. It is therefore difficult to make firm conclusions and recommendations have to be given cautiously. It may be appropriate, however, to give certain guidelines.

It appears that even very mild processing conditions such as cooking at 100°C for less than half an hour may cause a slight reduction in protein quality. Since some kind of processing is required for storage and transport, in addition to being part of a cultural tradition, the question is not to avoid processing but to optimise the conditions under which they are done.

Cooking at 95–100°C reduces protein and amino acid digestibility by a few per cent. In addition some soluble proteins, small peptides and free amino acids are lost in the cooking water. From a nutritional point of view, there is no reason to promote cooking of fish. When economical, technical and hygiene conditions allow cooking at temperatures lower than 100°C (i.e. at lower pressure) this practice should be encouraged.

Addition of salt may be necessary to preserve the fish. Salt levels should be kept at the lowest possible level to avoid harmful medical effects and possible negative effects on protein utilisation. Whenever possible the addition of nitrate should be avoided since apart from being toxicologically undesirable it may reduce protein quality.

Drying (dehydration) *per se* does not have an adverse effect on protein utilisation. Drying may enhance oxidation and rancidity, however, and thereby cause a slight reduction in protein quality. The degree to which drying adversely affects protein quality is dependent on drying temperature, and possibly drying time. Whenever possible drying should be conducted at 70 to 80°C or lower at which temperature damage to protein quality is negligible. Drying at higher temperatures (i.e. 115°C or higher) should be avoided due to profound negative effects on protein quality. Judged from protein quality, whenever there is a choice between processing temperature and processing time, reduction in temperature (down to 70°C) should have priority over a reduction in processing time. This recommendation does not include certain types of modern dryers not commonly used in the processing of fish (notably spray dryers), in which drying is completed within a fraction of a minute.

Evidence that the effects of temperature during drying are different from those during smoking is not available. Smoke contains carbonyls which react with lysine and reduce protein quality, the more smoke the greater the effect. From a protein quality point of view there is no reason to encourage smoking of fish. Whenever smoking is used, the length of smoking and the concentration of smoke should be kept to the lowest possible level.

REFERENCES

Adachi, R.R., Sheffner, L. and Spector, H. (1958). The *in vitro* digestibility and nutritional quality of dehydrated beef, fish and beans. *Food Res.* **23**, 401–5.

Aitken, A. and Connell, J.J. (1979). Fish. In: *Effects of Heating on Foodstuffs*, R.J. Priestley (Ed.), Applied Science Publishers, London, pp 219–54.

Aitken, A., Jason, A.C., Olley, J. and Payne, P.R. (1967). Effect of drying, salting and high temperatures on the nutritive value of dried cod. *Fishing News Intern.* **6**(9), 42–3.

Baba, H. (1960). On the available lysine in protein of fish and algae. *Bull Japan Soc. Sci. Fisheries* **26**, 330–3.

Bender, A.E. (1972). Processing damage to protein food. A review. *J. Food Technol.* **7**, 239–50.

Bender, A.E. (1977). The effect of heat on protein rich foods. In: *Food Quality and Nutrition*, W.K. Downey (Ed.), Applied Science Publishers, London, pp. 411–26.

Bjarnarson, J. and Carpenter, K.J. (1970). Mechanisms of heat damage in proteins 2. Chemical changes in pure proteins. *Brit. J. Nutr.* **24**, 313–29.

Buttkus, H.J. (1970). Accelerated denaturation of myosin in frozen solution. *J. Food Sci.* **35**, 558–62.

Carpenter, K.J. (1973). Damage to lysine in food processing: its measurement and its significance. *Nutr. Abstr. Rev.* **43**, 423–51.

Carpenter, K.J., Lea, C.H. and Parr, L.J. (1963). Chemical and nutritional changes in stored herring meal 4. Nutritional significance of oxidation of the oil. *Brit. J. Nutr.* **17**, 151–69.

Cesare, L.F.Di. (1980). Effetto dei trattamenti termici sulla lisina utilizzabile nelle proteine della sardina e dei fagioli. (The effect of heat treatment on the available lysine of the proteins of sardines and beans.) *Riv. Ital. Sostanze Grasse* **57**, 91–4.

Connell, J.J. (1962). Fish muscle proteins. In: *Recent Advances in Food Science*, J. Hawthorn and J.M. Leitch (Eds), Butterworths, London, pp. 136–46.

Cutting, C. L. (1962). The influence of drying, salting and smoking on the nutritive value of fish. In: *Fish in Nutrition*, E. Heen and R. Kreuzer (Eds), Fishing News (Books), London, pp. 161–79.

De Groot, A.P. (1963). The influence of dehydration of foods on the digestibility and biological value of the protein. *Food Technol.* **17**(3), 103–7.

El-Wakeil, F.A., Khairy, M., Morsi, S. and Abo-Raya, S.H. (1975). Chemical and biological evaluation of fresh and preserved Lake Nasser's fish. II Salt-dried keshr-bayad *Lates niloticus*. *Egypt. J. Food Sci.* **3**, 65–72.

Gardner, H.W. (1979). Lipid hydroperoxide reactivity with proteins and amino acids. A review. *J. Agr. Food Chem.* **27**, 220–9.

Herborg, L., Vilien, F., Bruun, A. and Eggum, B. (1974). Roller-dried fish protein. In: *Fishery Products*, R. Kreuzer (Ed.), FAO, Rome, pp. 309–11.

Hoffman, A., Barranco, A., Francis, B.J. and Disney, J.G. (1977). The effect of processing and storage upon the nutritive value of smoked fish from Africa *Tropical Sci.* **19**, 41–53.

Kennedy, T.S. and Ley, F.J. (1971). Studies on the combined effect of gamma radiation and cooking on the nutritional value of fish. *J. Sci. Food Agr.* **22**, 146–8.

Konosu, A. and Hashimoto, Y. (1959). Change of free amino acids during the manufacturing process of 'Katsuwobushi' (dried bonito). *Bull. Japan Soc. Sci. Fisheries* **25**, 307–11.

Konosu, S. and Matsuura, F. (1962). Tryptophan content of fish meat with special reference to the protein score. In: *Fish in Nutrition*, E. Heen and R. Kreuzer (Eds), Fishing News (Books), London, pp. 261–2.

Krauze, S., Oledska, R. and Buczynska, Z. (1970). Trawienie enzymatyczne in vitro bialka zwierzecego cz.II. Badanie strawnasci ryb surowych poddanych procesom termicznym (Enzymatic digestion *in vitro* of animal proteins. II Investigation of digestibility of raw fish as well as fish subjected to thermal treatment.) *Roczniki Panstwowego Zakladu Hig.* **21**, 287–92.

Ledward, D.A. (1979). Proteins. In: *Effects of Heating on Foodstuffs*, R.J. Priestley (Ed.), Applied Science Publishers, London, pp. 1–34.

Lee, K.-H. and Ryu, H.-S. (1987). Evaluation of seafood protein quality as predicted by C-PER assays. In: *Seafood Quality Determination*, D.E. Kramer and J. Liston (Eds.), Elsevier Science Publishers, Amsterdam, pp. 473–85.

Lionarons, H. (1970). Preservation of fish and fishery products. *Voeding* **31**, 525–32.

Maekawa, A., Isugi, M. and Matsumoto, K. (1975). Nutritive value of frozen Antarctic krills. *New Food Ind.* **17**, 25–30.

Munro, I.C. and Morrison, A.B. (1965). Effects of salting and smoking on protein quality of cod. *J. Fisheries Res. Board Can.* **22**, 13–16.

Mustafa, E.M. (1966). Beitrag zur Qualitätsbeurteilung von Fischprotein unter besonderer Berucksichtigung der Qualitätsveränderung durch die Trochnungstemperatur. (Contribution to the determination of quality of fish proteins with specific reference to variations in quality caused by drying temperature.) Dissertation zur Erlangung des Doktorgrades der Landwirtschaftlichen Fakultät der Justus Liebig, Universität Giessen, 202 pp.

Nagayama, F. (1960). Studies on the browning of fish flesh – II. Change of sugar content by heat process and browning. *Bull. Japan Soc. Sci. Fisheries* **26**, 1026–31.

Nagayama, F., Hiraide, H. and Sano, K. (1962). Studies on the browning of fish flesh – XI. Effect of heat on the levels of sugar, phosphate, arginine, and lysine in flesh. *Bull. Japan Soc. Sci. Fisheries* **28**, 1188–91.

Njaa, L.R., Utne, F. and Braekkan, O.R. (1968). Protein value of cod and coalfish and some products for the young rat. *Report Technol. Res. Norweg. Ind.* **5**(4).

Obileye, T. and Spinelli, J. (1978). A smoked minced tilapia product with enhanced keeping qualities. In: *Symposium on Fish Utilization Technology and Marketing in the IPFC Region*, 18th session, Manilla, Philippines and Bangkok, FAO, pp. 242–7.

Opstvedt, J., Miller, R., Hardy, R.W. and Spinelli, J. (1984). Heat-induced changes in sulfhydryl groups and disulfide bonds in fish protein and their effect on protein and amino acid digestibility in rainbow trout (*Salmo gairdneri*). *J. Agr. Food Chem.* **32**, 929–35.

Opstvedt, J., Søbstad, G. and Hansen, P. (1987). Fish protein concentrates for preruminant calves: effects of processing temperatures and source of fish on protein digestibility and biological value. *Animal Feed Sci. Technol.* **18**, 181–96.

Proctor, B.E. and Lahiry, N.L. (1956). Evaluation of amino acids in fish processed by various methods. *Food Res.* **21**, 91–2.

Sawant, P.L. and Magar, N.G. (1961). Nutritive value of canned fish: effects of canning, storage and antioxidants. *J. Sci. Ind. Res.* (India) **20D**, 313–16.

Seet, S.T. and Brown, W.D. (1983). Nutritional quality of raw, precooked and canned albacore tuna (*Thunnus alalunga*). *J. Food Sci.* **48**, 288–9.

Sheikh, A-S. and Shah, F.H., (1974). Effect of heat on the *in vitro* digestibility of fish protein. *Pakistan J. Sci. Ind. Res.* **17**, 136–8.

Shinno, H., Kisaka, M. and Nakamura, S. (1965). Amino acids in muscle proteins of flat fish and digestibility by pepsin. *Report Food Kyoto Women's Coll.* **17**, 12–18.

Shorrock, C. and Ford, J.E. (1978). Metabolism of heat-damaged proteins in the rat. Inhibition of amino acid uptake by 'unavailable peptides' isolated from enzymatic digests of heat-damaged cod fillet. *Brit. J. Nutr.* **40**, 185–91.

Tahiri-Zagret, M. (1969). Composition et valeur alimentaire de certains poissons fumes-seches de Cote D'Ivoire (Composition and nutritional value of certain smoke-dried fish of the Ivory Coast). *J. West African Sci. Assoc.* **14**, 161–80.

Thomas, M.H. and Calloway, D.H. (1961). Nutritional value of dehydrated foods. *J. Am. Dietet. Assoc.* **39**, 105–16.

Umoh, I.B., Ayalogu, E.O. and Bassir, O. (1980). Evaluation of the nutritive value of some lesser known protein sources in Nigerian peasant diets. *Ecol. Food. Nutr.* **9**, 81–5.

Wilder, O.H.M. and Kraybill, H.R. (1947). Effect of cooking and curing on lysine content of pork luncheon meat. *J. Nutr.* **33**, 235–42.

Yabuki, Y. and Izumi, K. (1966). On the amino acid contents in the hot water extracts from food materials (Part 4). On the amino acid contents in the hot water extracts from the meats of fishes and shell-fishes. *J. Japan Soc. Food Nutr.* **8**, 357–8.

Yanes, E., Ballester, D. and Donoso, G. (1970). Effect of drying temperature on quality of fish protein. *J. Sci. Food Agr.* **21**, 426–8.

Yasui, A. (1974). Nutritive value of bonito meat. *J. Home Econ.* (Tokyo) **25**, 265–372.

Yoshikawa, Y. and Tamamoto, I. (1942). Change in chemical compositions of fish during smoking process-I. Cold smoked herring. *J. Fisheries* **50**, 6–11.

Ziemba, Z. (1967). Nonenzymatic browning reactions in smoked foods. *J. Nutr. Dietet.* **4**, 122–9.

APPENDIX 1

EXPLANATIONS OF TERMS FOR PROTEIN QUALITY USED IN CHAPTER 2

***In vitro* or Laboratory Tests**

In vitro methods for protein quality evaluation refer to tests using chemical compounds and/or mono-cellular organisms to predict protein quality in higher animals. These methods will at the best give relative values, and figures obtained thereby should not be applied directly to higher animals.

FDNB (1-fluoro-2,4-dinitro-benzene)-reactive lysine measures the concentration of free epsilon amino groups in a protein and indicates the concentration of unbound lysine. The results as relative values are generally in good agreement with values of biologically available lysine determined in higher animals.

In vitro digestibility uses protein hydrolysis by a single enzyme, or a combination of two or more proteolytic enzymes, to indicate protein digestibility in higher animals. The values obtained depend on the method employed and in particular on the concentration of enzyme used and should be used only as relative figures.

Pepsin digestibility uses pepsin as the sole proteolytic enzyme. Different methods applying different pepsin concentrations are in use, leading to widely different values for protein hydrolyses.

Pepsin digest residue index (PDR) compares the release of 11 essential amino acids from a test protein with that of casein.

Relative nutritive value (RNV) uses the growth of a single-cell organism (bacterium or *Tetrahymena*) on a test protein compared with that on casein taken as 100.

Trypsin indigestible substrate (TIS) assesses *in vitro* digestibility using the single proteolytic enzyme, trypsin, by measuring the amount of insoluble material left (mg/g) after incubation.

In vivo or Animal Tests

Apparent protein digestibility (PDA), otherwise known as digestible crude protein (DCP), measures the difference between intake and faecal excretion of protein, given as a percentage of protein intake:

$$\text{PDA} = \frac{(\text{dietary protein intake} - \text{faecal protein excretion}) \times 100}{\text{dietary protein intake.}}$$

True protein digestibility (TPD), otherwise known as true digestible crude protein (TDCP), measures the amount of protein absorbed during digestion, taking into account the excretion of nitrogenous compounds into the digestive tract:

$$\text{TPD} = \frac{[\text{dietary protein intake} - (\text{faecal protein} - \text{metabolic protein})] \times 100}{\text{dietary protein intake.}}$$

TPD will, for the same protein sample, give higher values than PDA.

Nitrogen balance (NB) measures the difference between nitrogen intake and total nitrogen excretion as a percentage of nitrogen intake:

$$\text{NB} = \frac{[\text{N intake} - (\text{N in faeces} + \text{N excreted in urine})] \times 100}{\text{N intake.}}$$

Biological value (BV) measures the balance between nitrogen intake and total nitrogen excretion, taking into account the unavoidable metabolic nitrogen losses:

$$\text{BV} = \frac{\begin{array}{c}\text{N intake} - [(\text{total faecal N} - \text{metabolic faecal N}) \\ + (\text{total urinary N} - \text{endogenous urinary N})]\end{array}}{\text{N intake} - (\text{total faecal N} - \text{metabolic faecal N}).} \times 100$$

Net protein utilisation (NPU) measures the utilisation of a test protein for body protein synthesis:

$$\text{NPU} = \frac{\text{carcass protein gain}}{\text{dietary protein intake}} \times 100.$$

NB, BV and NPU measure, in addition to digestibility, also the utilisation of the absorbed protein for body protein synthesis.

Protein efficiency ratio (PER) relates weight gain to dietary protein intake using:

$$\text{PER} = \frac{\text{live weight gain}}{\text{dietary protein intake.}}$$

Fasting amino acid levels (PAA) measures the concentration of systemic plasma amino acids which will reflect dietary protein quality.

Gross protein value (GPV) measures the effect on growth of a test protein relative to a reference protein (usually casein) when either is added to a basal ration which usually consists of cereals:

$$\text{GPV} = \frac{(\text{weight gain on test protein} - \text{weight gain on basal ration})}{\text{weight gain on reference protein} - \text{weight gain on basal ration.}} \times 100$$

Chapter 3

EFFECTS OF DRYING AND SMOKING ON LIPIDS OF FISH

E. G. BLIGH, S. J. SHAW AND A. D. WOYEWODA
Technical University of Nova Scotia, Nova Scotia, Canada

INTRODUCTION

The traditional processes of drying and smoking of fish have been passed down by our ancestors and in some aspects have not changed substantially. Improvements, nevertheless, have been made in equipment and formulation.

In the early days, the major goals of smoking and drying were to preserve perishable seafoods from times of plenty to less bountiful times and also to enable broader distribution to areas removed from the sea. Negative effects on the nutritive value of the consumer products were considered insignificant in light of the overall benefits provided.

In today's modern world, refrigeration, cold storage and thermal processing have made traditional products more delicacies than essentials. Streamlined processes, however, sometimes orientated more toward simple flavour and texture enhancement rather than preservation, can have undesirable side-effects (Gilbert and Knowles, 1975).

The purpose of this paper is to examine the lipid component of fish and to explore some of the pertinent scientific literature with regard to the effect of drying and smoking on lipids.

LIPIDS

The chemical composition of fish lipids differs from that of other naturally occurring fats and oils in: (1) containing a greater proportion of highly unsaturated fatty acids; (2) possessing larger quantities of fatty acids with chain lengths greater than 18 carbons; (3) having polyunsaturates, primarily of the n-3 family rather than the n-6 family; and (4) in general, possessing a greater variety of lipid compounds.

The lipid content of fish flesh ranges from less than 1 % to 22 % of the main constituents in different species, and varies more widely than protein or moisture. Major fluctuations are seen with location, season, diet, exercise and maturity. Usually, the lipid plus water content of fish flesh equals 80 %; i.e. as fat content increases, water content decreases correspondingly.

Several classes of lipids may be identified in marine fish flesh. Simple lipids (i.e. fats, oils) or 'triglycerides' are esters of glycerol with monocarboxylic or fatty acids (triacylglycerols). Of the fatty acids in marine triglycerides, only 25 to 40 % are saturated; the remainder are either monounsaturated or polyunsaturated. Unique properties of marine oils can usually be attributed to their polyunsaturation. Wax esters (long-chain alcohols esterified with fatty acids), sterols, free alcohols, and hydrocarbons comprise another class of lipids, accounting for approximately 1 % of total lipid in most edible species. Compound lipids include phospholipids, glycolipids and lipoproteins. In association too are the fat soluble vitamins A, D, E and K.

Marine lipids are present as either depot or non-depot fats. The location and quantities of depot lipids vary greatly between species and with season. These fats serve as an energy reserve especially for the production of eggs and milt (Lovern, 1962; Ackman, 1980). In lean fish (e.g. cod, hake) depot lipids are located in the liver and peritoneum. For fatty fish (e.g. herring, mackerel, salmon) they are found as extracellular deposits in the muscle, in the mesentery (Labuza, 1971) or subdermally (Ackman, 1980). Although depot lipids consist primarily of triglycerides, in elasmobranchs (e.g. sharks, rays) lipids other than triglycerides may be present. Non-depot or structural (e.g. cell wall) lipids include phospholipids, sphingolipids and sterols. These are not readily extractable and considerable variation exists in concentration in different parts of the muscle. Richer deposits of total lipid with an increased concentration of haem pigments (Labuza, 1971) are found in dark muscle, located along the lateral line or in the flank, or as red muscle (Lovern, 1962). Belly flaps and tail muscle may also contain more lipid than head or centre sections (Ackman, 1980).

The high degree of unsaturation, in the form of multiple double bonds in the fatty acids, renders fish lipids highly susceptible to the development of oxidative rancidity. Attack by molecular oxygen on unsaturated fatty acids, proceeds by a free radical mechanism and is characterised by a slow initiation period, followed by an accelerating rate of oxygen absorption with formation of hydroperoxides, which eventually break down to carbonyls, and polymerised products (Olcott, 1962). These intermediary compounds, relatively unstable, cause the oxidation of pigments, loss of vitamins and objectionable flavours which are characteristic of 'rancid' oil (Khayat and Schwall, 1983). Termination of free radical reactions by a natural scavenger is the usual final step in the chain mechanism.

Oxidative rancidity is most often measured by the 2-thiobarbituric acid (TBA) test, peroxide value (POV), or carbonyl value (COV) (Gray, 1978; Melton, 1983). The transient nature of the products being measured often necessitates the use of at least two tests for oxidative quality determination. Inoue (1970) reported that POV was not a good index of rancidity during storage of dried anchovy but that TBA gave a better correlation with overall quality.

Lipid hydrolysis in seafoods is catalysed by lipases which cleave fatty acids from the glycerol backbone of both triglycerides and phospholipids (Hardy, 1980). In the case of triglycerides, the reaction usually proceeds to monoglyceride plus two free fatty acids (FFA). Lipid hydrolytic activity is determined by measuring the FFA level in the total extracted lipid. Lipid hydrolysis by itself has no nutritional significance, but the accumulation of FFA in fish oils is undesirable due to the secondary reactions catalysed, such as increased susceptibility to oxidation and the consequent development of off-flavours (Lovern, 1962). There has also been a long debate over claims that lipid hydrolysis and the associated accumulation of FFA contributes to accelerated protein denaturation in frozen fish (see Sikorski *et al.*, 1976; Matsumoto, 1979).

Determinant factors to oxidative susceptibility of fish lipids, in addition to their unsaturation, include: (1) distribution of fat within the body (belly flaps and tail muscle are more susceptible); (2) contact with pro-oxidants in skin, dark muscle or aqueous components (e.g. haem pigments; Ke *et al.*, 1977; Woolfe, 1975); (3) presence or absence of other chemical compounds (e.g. sodium chloride; Castell *et al.*, 1965); and (4) external factors such as heat, light, UV rays and available oxygen (Khayat and Schwall, 1983).

For example, Isa (1961) found that the head and visceral fats of anchovy were more susceptible to oxidation than the muscle fats and that the former, particularly the visceral fats, varied greatly with season and sex.

Currently, special interest has been focused on the n-3 fatty acids extensively found in fish oils. Clinical evidence has shown that these polyunsaturated fatty acids can have beneficial effects on human health and may reduce cardiovascular disease and rheumatoid arthritis (Lands, 1986; Eichner, 1987; Kinsella, 1987). Inasmuch as these fatty acids (e.g. eicosapentaenoic acid, 20:5 n-3, and docosahexaenoic acid, 22:6 n-3) are highly unsaturated they are particularly prone to oxidation and thus might readily lose their nutritional and clinical effectiveness when subjected to oxidising conditions during the drying and smoking of fish and the storage of dried and smoked products.

DRYING

Effect of Heating

The basis for preservation by drying is removal of sufficient quantities of water to lower water activity (Aw) to a level which inhibits bacterial growth and slows chemical and enzymatic reactions. Some form of heat is usually involved, ranging from the sun in outdoor drying, to the relatively severe heat of a commercial dryer.

Heating undoubtedly causes oxidation of lipids in fish (Aitken and Connell, 1979). Perhaps the major impact of this oxidation lies in the potential loss of available lysine. Carbonyl compounds from lipid oxidation, in the presence of mild heat, can react with the ϵ-amino group of lysine in a manner similar to that of reducing sugars to complete the Maillard reaction. Cutting (1962) reported the loss of biologically available lysine (and histidine) due to reaction of oxidation products with basic nitrogenous constituents in fish flesh. He indicated that this Maillard reaction, in combination with oxypolymerisation of unsaturated fish oils, was responsible for darkening, toughening, and the unpalatable, bitter flavour associated with lengthy drying processes and storage of dehydrated fish.

With excessive heating (140–150°C), reductions in available methionine, cystine and tryptophan are also possible (FAO, 1981). Takama and Zama (1975) observed that lactones were formed when the lipids of rainbow trout were exposed to 125°C for 24 h. Arai (1980) studied the effect of lipid oxidation on nutritive value of protein, supplied as defatted powders of boiled and dried anchovies, both with and without antioxidant. Butylated hydroxyanisole (BHA) and vitamin E effectively inhibit lipid oxidation. From a storage experiment on squid and mackerel, Ikegami *et al.* (1971) concluded that lipid oxidation and protein denaturation resulted in a decrease in nutritive value.

Effect of Moisture

Koizumi *et al.* (1980) compared rates of lipid oxidation in salted freeze-dried bluefin tuna (red muscle fish) and sablefish (white muscle fish) as a function of moisture equilibrium relative humidity, in order to find the optimum storage conditions for salted dried fish. From lipid composition and POV, it is evident that at relative humidities between 0 and 71%, salted freeze-dried bluefin tuna containing 19% lipid on a dry basis undergo considerable triglyceride oxidation with an increase in the polar lipid fraction, especially in the first 7 days of storage at 25°C. The phospholipid increase was attributed to non-separation of triglyceride oxidation products from phospholipids by the chromatographic technique used. Oxidation proceeds more rapidly at lower relative humidity and changes in fatty acid composition were noted. At 0% and 11% relative humidities, polyenoic acids are completely oxidised. At 52% and 71% relative humidities, however, the polyenoic acids although lower, appear more stable in all lipid classes, i.e. phospholipids, triglycerides and free fatty acids.

Salted freeze-dried sablefish with 49·1% lipids on dry basis and polyenoic acids in similar proportion, give different results (Koizumi *et al.*, 1980). At 11% relative humidity, the percentage of triglycerides decreases slowly with an accompanying gradual phospholipid increase during 42 days' storage. Peroxide values confirm a much slower oxidation of triglyceride at 11% relative humidity compared to tuna and changes in the percentages of total polyenoic and monoenoic acids were not observed. At 71% relative humidity, a decreased triglyceride content is attributed to hydrolysis by an endogenous enzyme system in the meat rather than oxidation since POV is low and there had been some FFA increase.

The authors concluded that sablefish is much more stable to lipid oxidation than bluefin tuna under salted freeze-dried conditions. The difference is not due to fatty acid composition, since both fish contain polyenoic fatty acids in similar amounts in their lipids. Rather, the significant difference between the fish is the content of haemoprotein in the flesh with tuna having abundant levels and sablefish being low. Other workers have also reported haemoglobin as a pro-oxidant (Woolfe, 1975).

Koizumi *et al.* (1978) studied the effect of relative humidity on the rate of lipid deterioration of freeze-dried big-eye tuna and halibut stored at 25°C. Results show a definite correlation. At low relative humidity, lipids of both species are oxidised although at a faster rate in tuna than halibut.

The triglyceride content of halibut hydrolyses enzymatically even at 11% relative humidity, while enzymatic hydrolysis of tuna triglyceride was not detected. At higher relative humidities (52% and 71%) lipids from neither fish undergo extensive oxidative deterioration during the experiment.

Traditional Products

Rao and Bandyopadhyay (1983) studied lipid composition of commercial samples of salted sun-dried mackerel. Compared to fresh fish, these mackerel have a decreased 'glyceride' content and higher FFA indicating hydrolytic cleavage. Oxidative breakdown is also evident from low iodine values and high POV and TBA values. The authors concluded that sun-drying causes considerable loss of polyunsaturated fatty acids.

Howgate and Ahmed (1972) studied heating and dehydration of cod (lean fish from cold waters) and hilsa (fatty fish from tropical waters). After drying in a mechanical dryer at 30°C, considerable oxidation in the fat of both species was measured by POV and malonaldehyde content. After storage a higher rate of oxidation was reported in unwrapped dried samples due to greater oxygen availability. Miwa (1963) found that boiling caused little hydrolysis of phospholipids in Alaska pollack, Atka mackerel and flat fish but that FFA increased during subsequent processes.

Lovern (1962) reported a progressive decline of lipid hydrolysis in salted cod with decreasing moisture content. Hydrolysis in Gaspé cured cod (lightly salted) is continual throughout curing, while most activity occurs in the early stages with a heavy cure. Hydrolysis ceases when lipids contain 50% FFA, probably due to lipase inactivation. Cutting (1962) also reported lipolysis during drying of heavily salted and Gaspé cured cod, with FFA reaching 50% before enzyme inactivation. He further commented that cutting and pressing damage during traditional curing lead to mechanical release of oils, which are spread over the cut surface, exposed to oxygen, and destined to oxidative rancidity.

In traditionally cured products, hydrolytic breakdown of fish lipids is not totally unacceptable, since it imparts desirable flavours. However, the contribution of oxidation remains unclear. In some places (e.g. Hong Kong) the dark colour and rancid odour of cured fish is a sign of quality. With dried salted cod, however, a dark colour is undesirable. Therefore, the use of salt containing less than 0·2 ppm copper is recommended since cupric ions will activate browning reactions (Dyer and Gunnarson, 1954; Shewan, 1955).

Fish Meal

While there have been long debates concerning the detrimental effects of drying (heating) on the nutritional value of fish, the most definitive studies have been done with fish meals. Inasmuch as the drying conditions used in the manufacture of fish meal may sometimes be considered severe in comparison with those employed in seafood processing, it is reasonable to conclude that any effects on nutrients would be greater in fish meal than in other seafoods. It must be emphasised, however, that fish meals are used in compounded animal feedstuffs to make good nutrient deficiencies which might otherwise be present and their major contributions are to improve the quality of the dietary protein.

Heating and oxidation are major considerations in the production and storage of fish meals especially in those produced from fatty fish such as herring, sardine and menhaden (Olcott, 1962). March *et al.* (1961) reported that herring meal oxidises more quickly at −20°C than at 25°C. They endeavoured to explain this unusual finding; decreases in iodine number and in the amount of lipid extractable at both temperatures were noted. Antioxidant treatment prevents the changes. Labuza (1971) stated that rapid oxidation of dehydrated fish meal could induce spontaneous combustion. Intimate contact between the meal and fish blood, bones, and scales during processing has been known to set up the reaction. If moisture content is held below 6%, the meal will be more stable. Excessive heating (140–150°C) can lead to reductions in available lysine, methionine, cystine and tryptophan, due to interactions with oxidised oil (FAO, 1981) as well as the loss of available lysine due to the reaction of carbonyl components with the ϵ-amino groups of fish proteins (Olcott, 1962).

Tarr *et al.* (1954) reported that providing the diet contained adequate vitamin B complex, the nutritional value of herring meal produced by direct flame drying was not significantly different from that of vacuum-dried meal in which the drying temperature did not exceed 43°C. The heating of vacuum-dried meal at 149°C for 3 h, however, causes severe impairment of nutritional value due to the loss of essential amino acids (Bissett and Tarr, 1954). These losses are not encountered when meal is heated for periods of up to 1 h at 149°C. Additional studies pertaining to the lipid component of heated fish meals were carried out to resolve questions related to possible toxic effects and poor digestibility due to fat oxidation and polymerisation. Chicken feeding studies showed that changes in the oil fraction of fish meal does not influence the nutritional value providing rations are adequately supplemented with vitamins (Biely *et al.*, 1955). Subsequent trials demonstrated that the animals' ability to utilise herring meal lipids decreases with storage and oxidation but is improved by incorporation of BHT (butylated hydroxytoluene) as antioxidant (March *et al.*, 1962).

SMOKING

The preservative process of smoking combines the effects of salting, drying, heating and smoking. The first step, salting, usually involves soaking in a prepared salt solution to firm the flesh and impart flavour. Subsequent drying, often performed outdoors or occasionally in a smokehouse, removes excess water from the flesh. A combined heating/drying/smoking process entails application of smoke, at temperatures of about 30°C for a cold smoked product or alternatively 70 to 80°C for a hot smoked product. Cold smoking does not cook the flesh nor coagulate the protein, so refrigerated storage and cooking are necessary before consumption (smoked salmon is an exception). The hot smoking process adequately cooks the flesh. The preserving effect of smoking on fishery products is credited to a combination of surface drying, salting, and deposition of antioxidant (phenolic) and antimicrobial constituents on the fish (Gilbert and Knowles, 1975).

Relatively little has been published on smoke-induced changes in fish lipids. If the processes involved are considered separately, some indication of the probable effects on lipids may be seen. Castell *et al.* (1965) reported that salt accelerated rancidity development in blended cod at 0°C, possibly due to the presence of trace heavy metals in the salt, and that addition of chelating agents and commercial antioxidants retards the process.

Koizumi *et al.* (1980) found that dried-salted fish products are more susceptible to oxidative changes than freeze-dried flesh, because the added salt functions as a pro-oxidant. Hobbs (1982) concluded that the oxidising effects of salt on fish lipids are outweighed by the antioxidant effect of smoke constituents. Cutting (1962) stated that although salt promoted rancidity in frozen fatty fish, where cured fish are covered with a saturated salt brine, oxygen is rendered almost insoluble and rancidity develops extremely slowly. He concluded that the effect of salt as a pro-oxidant for fish oils is not clear.

Other evidence exists that salt has a protective effect on lipids. Nambudiry (1980) stated that it is a pro-oxidant in fatty species at lower concentrations, but as concentrations increase, salt inhibits lipid oxidation. In addition to its effect on oxidation, salting also exerts an osmotic effect on fish flesh; thus, oil and other constituents may be lost from fatty fish (Cutting, 1962).

In a comprehensive study on the effects of smoking on some functional properties of spring and autumn Atlantic mackerel (*Scomber scombrus*), Bhuiyan *et al.* (1986) reported that oxidative deterioration results from the salting, heating, smoking and exposure to oxygen to which the fish are subjected. Thiobarbituric acid (TBA) values for smoked mackerel are almost double those of unsmoked material (6·86 vs 3·98 μmol/kg fish) and peroxide values are also markedly increased (for unsmoked, 1·30 vs 8·96 meq/kg fat for smoked). Salt is considered to be a major contributing factor.

Analysing major lipid classes of Atlantic mackerel before and after salting, mild heating and smoking, Bhuiyan *et al.* (1986) found that overall the process has an insignificant effect on the two major muscle lipid classes, triglycerides and phospholipids. Additionally, significant changes were not detected in overall fatty acid composition (saturated, monoenoic and polyunsaturated fatty acids). Free fatty acids (FFA) were at 0·1 % or less, indicating little hydrolytic breakdown in the smoked mackerel.

Phenolic substances in wood smoke are often credited with providing some protection against oxidation (FAO, 1981), while other authors have concluded that smoking has no effect. Early work by Takahashi and Masuda (1938) demonstrated that herring oil exposed to smoke and subsequently held at 40°C for up to 35 days showed little increase in weight due to the uptake of oxygen in contrast to a marked weight increase for non-smoked oil. They also reported that the POV for the smoked herring oil was negligible when compared with that for the fresh herring oil. Similar antioxidant effects of wood smoke were observed by Ohtani (1938) when he exposed sardine oil to smoke for 1 h and monitored the development of oxidative rancidity over a period of 16 days. Fretheim *et al.* (1980) also reported that smoke condensates have antioxidant effects, even at low concentrations. Yamamura (1951) found only slight changes in iodine value and peroxide value in smoked shark meat after storage for 3 months. A paper by Kinoshita (1960) reported that butylated hydroxytoluene and hydroquinone were detected in smoked octopus. Woolfe (1975), however, found no evidence that smoking decreases the rate of oxidation in *Sardinella aurita* lipids where the fish have been smoked whole and the skin acts as a barrier to penetration of the flesh by smoke constituents. Peroxide values indicated that oxidation of the lipids is initiated by the drying portion of the process. Variation of temperature in the range of 60 to 100°C has little effect on oxidation rate. Moisture content, however, is important, At low moisture levels, a direct relationship exists between temperature and oxidation rate of the smoked product while at higher moistures, the opposite is true.

CONCLUSION

It appears that the preservative processes of smoking and drying serve to catalyse degradative changes in marine lipids. Smoking, being the less severe of the two, does not cause as extensive deterioration as drying. The major effect of lipid degradation is development of aesthetically unappealing odours and flavours while the effect of oxidative rancidity on nutritive quality of fish, especially in terms of protein availability, is debatable.

REFERENCES

Ackman, R.G. (1980). Fish lipids, Part 1. In: *Advances in Fish Science and Technology*, J.J. Connell (Ed.), Fishing News (Books), London, pp. 86–103.

Aitken, A. and Connell, J.J. (1979). In: *Effects of Heating on Foodstuffs*, R.J. Priestley (Ed.), Applied Science Publishers, London, pp. 219–54.

Aitken, A., Jason, A.C., Olley, J. and Payne, P.R. (1967). Effects of drying, salting, and high temperatures on the nutritive value of dried cod. *Fishing News Int.* **6**(9), 42–3.

Arai, K. (1980). Nutritive value of defatted powder from boiled and dried anchovy treated with antioxidants. *Bull. Tokai Reg. Fish. Res. Lab.* **102**, 25–9.

Bhuiyan, A.K.M., Ratnayake, W.M.N. and Ackman, R.G. (1986). Stability of lipids and polyunsaturated fatty acids during smoking of Atlantic mackerel (*Scomber scombrus*). *J. Am. Oil Chem. Soc.* **63**, 324–8.

Biely, J., March, B.E. and Tarr, H.L.A. (1955). The nutritive value of herring meals. 3. The effects of heat treatment and storage temperature as related to oil content. *Poultry Sci.* **34**, 1274–9.

Bissett, H.M. and Tarr, H.L.A. (1954). The nutritive value of herring meals. 2. Effect of heat on availability of amino acids. *Poultry Sci.* **33**, 250–4.

Castell, C.H., Maclean, J. and Moore, B. (1965). Rancidity in lean fish muscle. IV. Effect of sodium chloride and other salts. *J. Fisheries Res. Board Can.* **22**, 929–44.

Cutting, C.L. (1962). The influence of drying, salting and smoking on the nutritive value of fish. In: *Fish in Nutrition*, E. Heen, and R. Kreuzer (Eds), Fishing News (Books), London, pp. 161–79.

Dyer, W.J. and Gunnarson, G.K. (1954). Trace elements and yellow discoloration in salt fish. *Fisheries Res. Board Can. Prog. Rpts, Atlantic Coast Stations* **60**, 3–4.

Eichner, E.R. (1987). Health benefits from fish and fish oils. *Inter. Med. Special.* **8**(2), 49–52.

FAO (1981). *Prevention of Losses in Cured Fish.* Fisheries Technical Paper No. 219, FAO, Rome, 10 pp.
Fretheim, K., Granum, P.E. and Vold, E. (1980). Influence of generation temperature on the chemical composition, antioxidative and antimicrobial effects of wood smoke. *J. Food Sci.* **45**, 999–1002, 1007.
Gilbert, J. and Knowles, M.E. (1975). The chemistry of smoked foods: a review. *J. Food Technol.* **10**, 245–61.
Gray, J.I. (1978). Measurement of lipid oxidation: a review. *J. Am. Oil Chem. Soc.* **55**, 539–46.
Hardy, R. (1980). Fish lipids. Part 2. In: *Advances in Fish Science and Technology*, J.J. Connell (Ed.), Fishing News (Books), London, pp. 103–11.
Hobbs, G. (1982). Changes in fish after catching. In: *Fish Handling and Processing*, A. Aitken, I.M. Mackie, J.H. Merritt and M.L. Windsor (Eds), Torry Research Station, Ministry of Agriculture, Fisheries and Food, UK, pp. 20–7.
Howgate, P.F. and Ahmed, S.F. (1972). Chemical and bacteriological changes in fish muscle during heating and drying at 30°C. *J. Sci. Food Agric.* **23**, 615–27.
Ikegami, Y., Takai, Y. and Shibuya, K. (1971). Nutritive value of frozen food VIII. Nutritive value of frozen squid and mackerel. *Report. Inst. Nutr.* 1970, 73–8.
Inoue, T. (1970). Oxidation of oil contained in dried anchovy during storage. *Bull. Fac. Educ.* Kagawa Univ. **2**, 61–7.
Isa, Y. (1961). Studies on the boiled and dried anchovy III. On crude fat content among three body portions of head, viscera, and muscle and presence of the so-called oily anchovy. *Bull. Japan. Soc. Sci. Fisheries* **27**, 1080–3.
Ke, P.J., Ackman, R.G., Linke, B.A. and Nash, D.M. (1977). Differential lipid oxidation in various parts of frozen mackerel. *J. Food Technol.* **12**, 37–47.
Khayat, A. and Schwall, D. (1983). Lipid oxidation in seafood. *Food Technol.* **37**(7), 130–40.
Kinoshita, G. (1960). Effect of rancid oil on cooking IV. Chemical constituents of smoke and their action. *J. Japan. Soc. Food Nutr.* **13**, 250–6.
Kinsella, J.E. (1987). *Seafoods and Fish Oils in Human Health and Disease*, Marcel Dekker Inc., New York, 317 pp.
Koizumi, C., Tiyama, S., Wada, S. and Nonaka, J. (1978). Lipid deteriorations of freeze-dried fish meats at different equilibrium relative humidities. *Bull. Japan. Soc. Sci. Fisheries* **44**, 209–16.
Koizumi, C., Terashima, H., Wada, S. and Nonaka, J. (1980). Lipid oxidation of salted freeze-dried meats at different equilibrium relative humidities. *Bull. Japan. Soc. Sci. Fisheries* **46**, 871–7.
Labuza, T.P. (1971). Kinetics of lipid oxidation in foods. In: *CRC Critical Reviews in Food Technology*, CRC Press, Cleveland, Ohio, pp. 355–405.
Lands, W.E.M. (1986). Overall supply of n-3 fatty acids. In: *Fish and Human Health*, Academic Press, London, pp. 136–48.
Lovern, J.A. (1962). The lipids of fish and changes occurring in them during processing and storage. In: *Fish in Nutrition*, E. Heen and R. Kreuzer (Eds), Fishing News (Books), London, pp. 86–111.

March, B.E., Biely, J., Goudie, C., Claggett, F. and Tarr, H.L.A. (1961). The effect of storage temperature and antioxidant treatment on the chemical and nutritive characteristics of herring meal. *J. Am. Oil Chem. Soc.* **38**, 80–4.

March, B.E., Biely, J., Claggett, F.E. and Tarr, H.L.A. (1962). Nutritional and chemical changes in the lipid fraction of herring meals with and without antioxidant treatment. *Poultry Sci.* **41**, 873–80.

Matsumoto, J.J. (1979). Denaturation of fish muscle proteins during frozen storage. In: *Behaviour of Proteins at Low Temperatures*, O. Fennema, (Ed.), American Chemical Society Washington, pp. 205–24.

Melton, S.L. (1983). Methodology for following lipid oxidation in muscle foods. *Food Technol.* **37**(7), 105–11, 116.

Miwa, K. (1963). Changes in phospholipids during handling and processing of fish. *Report Hokkaido Regional Fisheries Res. Inst.* **26**, 129–39.

Nambudiry, D.D. (1980). Lipid oxidation in fatty fish: the effect of salt content in the meat. *J. Food Sci. Technol.* **17**, 176–8.

Olcott, H.S. (1962). Oxidation of fish lipids. In: *Fish in Nutrition*, E. Heen and R. Kreuzer (Eds), Fishing News (Books), London, pp. 112–16.

Ohtani, T. (1938). Studies on fish meal-II. Discoloration of sardine oil added to absorbents. *Bull. Japan. Soc. Sci. Fisheries* **7**, 111–16, 217.

Rao, B.Y.K. and Bandyopadhyay, C. (1983). Lipid composition of salted sun-dried Indian mackerel (*Rastrelliger kanagurta*). *J. Food Sci. Technol.* **20**, 62–4.

Shewan, J.M. (1955). The browning of salt-cured white fish. *Food Manuf.* **30**, 200–3.

Sikorski, Z., Olley, J. and Kostuch, S. (1976). Protein changes in frozen fish. *Crit. Rev. Food Sci. Nutr.* **8**, 97–129.

Takahashi, E. and Masuda, G. (1938). Studies on the oil and fat in smoked fish-II. Nutritive value of smoked herring oil. *J. Japan. Agric. Chem.* **14**, 19–27.

Takama, K. and Zama, K. (1975). Lactones in heated fish lipid. *Bull. Fac. Fisheries*, Hokkaido Univ. **26**, 169–75.

Tarr, H.L.A., Biely, J. and March, B.E. (1954). The nutritive value of herring meals. 1. The effect of heat. *Poultry Sci.* **33**(2), 242–50.

Woolfe, M.L. (1975). The effect of smoking and drying on the lipids of West African herring (*Sardinella* spp.). *J. Food Technol.* **10**, 515–22.

Yamamura, Y. (1951). Preservative experiment of smoked shark fillet II. Change of oil components. *Bull. Japan. Soc. Sci. Fisheries* **16**, 510–12.

Chapter 4

THE EFFECT OF DRYING AND SMOKING ON THE VITAMIN CONTENT OF FISH

J. R. Burt
Torry Research Station, Aberdeen, UK

The flesh of fish contains small amounts of nearly all the vitamins with the exception of ascorbic acid, of which only insignificant amounts are present, and of vitamin D, which is present in substantial quantities in the flesh of fatty fish. As a consequence, unless a diet is based largely on fish, the stability of vitamins during processing will be of only minor importance (Cutting, 1962). Cutting (1962) in reviewing the literature then available made some comments and also reached a number of other conclusions which are still particularly germane. In general, figures for the vitamin contents of cured fish products were few, scattered and, as might be expected from seasonal and other causes of variability in composition of raw material, rather variable. Differences in processing technique would also introduce variations. This means that analyses of products, no matter how accurately conducted, could be unrepresentative and unreliable. Furthermore, the effects of processing cannot be deduced with any confidence by comparing analytical values for raw material and product unless such values relate to the same batches of material; comparisons of tabulated figures must be treated cautiously.

Since Cutting's (1962) review was published, various aspects of this field have been reviewed a few times (Daun, 1975; Bluestein and Labuza, 1975; Aitken and Connell, 1979; De Ritter, 1982) but few new data have been published. In the light of such a paucity of reliable information that is directly relevant, the approach to be taken in this section will be to survey the basic, stability properties of each vitamin in turn, to add what conclusions have already been drawn in their regard and, finally, to leave it to the nutritionist to consider what the likely effects of processes are going to be in the light of the respective processing régime (see Chapter 9) for any particular product. Taking the converse view, little work has been done on the kinetics of nutritional loss under conditions of constant temperatures and humidities but it is possible that process optimisation procedures could be calculated where necessary (Bluestein and Labuza, 1975).

Salting often precedes drying but is a customary prerequisite to smoking. The effect of low concentrations of salt on nutritional value have not been studied extensively but their influence is probably not significant (Daun, 1975). Salt does, however, exhibit pro-oxidant activity in fish flesh (Castell *et al.*, 1965; Tarr, 1969) and may increase the losses of some vitamins. The salting process itself may, however, induce a physical loss of water soluble constituents, including the B vitamins (Cutting, 1962), and it would be logical to assume that such losses as take place will be proportional to the losses of water occurring during the first stages of curing (Daun, 1975). Salting also induces losses of oil from fatty fish (Cutting, 1962).

Drying procedures may or may not involve exposure to heat but smoking usually does; in addition, for smoking, losses of nutrients may result from interaction with smoke components. Furthermore, any changes of nutritional value that take place because of dehydration could also be expected to happen when similar amounts of dehydration occur during smoking.

Smoked meat products exhibit an increased resistance to oxidative changes and, whilst oxidation can lead to the destruction of some vitamins (Daun, 1975), it would be expected that certain smoke constituents could help protect such oxidisable vitamins in smoked fish products. Since the particulate phase of smoke has greater antioxidant activity than the vapour phase (Tilgner *et al.*, 1965), the method of smoke generation may affect antioxidant activity within a smoked product (Tilgner and Daun, 1970).

FAT SOLUBLE VITAMINS

The fat soluble vitamins are generally more stable than the water soluble ones but they are prone to degradation at high temperatures in the presence of oxygen, perhaps by free radical mechanisms (Priestley, 1979). Fat soluble vitamins can interact with the products of lipid oxidation and measures used to protect lipids will also improve the retention of these compounds (Daun, 1975). The pro-oxidant action of salt and the antioxidant activity of some smoke constituents, already mentioned, are of relevance in this regard. Rather surprisingly data are not available on losses of fat soluble vitamins accompanying fat exudation during smoking (Daun, 1975) or associated with the losses of oil during the salting of fatty fish (Cutting, 1962). Losses of fat soluble vitamins occur during storage of salted fish and antioxidants have been used to help reduce these losses (Higashi, 1962).

Vitamin A

This vitamin is reported as being quite stable in food processing (Priestley, 1979) and losses of less than 10% have been observed during canning salmon, mackerel and swordfish (De Ritter, 1982). As a result of its unsaturated character, however, it is liable to become oxidised when heated in the presence of oxygen or exposed to ultraviolet light (Mejika, 1960; Bramsnaes, 1962; Priestley, 1979): this oxidation can be catalysed by copper and iron ions (Priestley, 1979). Losses can also occur in the presence of oxidised oils (Aylward and Haisman, 1969) and these phenomena account for reports that vitamin A is completely destroyed during the salting and drying of tropical products (van Veen, 1936). During smoking, and subsequent canning, little if any losses of vitamin A were reported (Cutting, 1962). Mejika (1960), on the other hand, found that losses by broiling fish were less than 20% compared with 25 to 40% by boiling in water.

Vitamin D

Vitamin D is also vulnerable to oxidation but is generally more stable than vitamin A to heat, ultraviolet light and oxygen (Priestley, 1979). It is normally stable during canning, provided air is excluded from the can (Bramsnaes, 1962) and, as with vitamin A, losses of it in herring flesh during smoking and subsequent canning are not appreciable (Cutting, 1962). Aitken and Connell (1979) summarising all the available data, came to the conclusion that vitamin D is only slightly reduced during cooking and canning.

Vitamin E

Vitamin E is quite heat resistant (Bunnell *et al.*, 1965; Priestley, 1979), but Bramsnaes (1962) reported that it is destroyed by heat during canning. It is highly susceptible to oxidation in the presence of oxygen and catalysts at high temperatures (Priestley, 1979) but, despite this susceptibility, Bluestein and Labuza (1975) failed to find any reports in the literature of losses of vitamin E during drying.

WATER SOLUBLE VITAMINS

In view of the diverse natures and chemical properties of the water soluble vitamins it is probably dangerous to treat them together in any respect other than their common solubility in water. It has, however, been stated (Priestley, 1979) that they are on the whole less stable than the fat soluble ones and that losses into processing water (where that is used) often far exceed losses due to thermal destruction. Despite these properties, Cutting (1962) concluded that losses of the B vitamins in mild smoking, hot or cold, are not appreciable even when the smoking is followed by canning and that mildly smoked fish is not detectably less nutritious than fresh, unprocessed fish. During conventional drying, losses of the water soluble vitamins are less than 10% (Bluestein and Labuza, 1975). Hard salted and dried fish, on the other hand have noticeably lower concentrations of the B vitamins.

A Japanese study on the B vitamins in marine products and their changes during processing and storage (Komata *et al.*, 1956, 1957) showed that losses of vitamins during heat sterilisation and successive storage were small and even thiamine was relatively stable. Large amounts of B vitamins were lost, however, during the pre-cooking treatment given to baby clams prior to canning and only 17 to 44% of the individual vitamins was retained throughout the whole procedure (Komata *et al.*, 1957). In the cases of canned mackerel and tuna, on the other hand, only thiamine showed appreciable losses (52 to 70%); over 80% of the amounts of certain other B vitamins present in the raw flesh remained after canning and storage (Komata *et al.*, 1956).

Thiamine

Of the B vitamins, thiamine is the most sensitive to degradation by heat and the most important factors governing its loss in heated foodstuffs are temperature, time of heating and pH; thiamine is more labile under alkaline than under acid conditions. Other factors which can affect its breakdown are the presence of metal ions, copper in particular, the form of the vitamin and the presence or absence of protective agents such as proteins, and this last point probably accounts for the fact that thiamine in natural products is more stable to heat than it is in either aqueous or buffer solution (Priestley, 1979).

So far as the stability of thiamine during processing of fish is concerned, Aitken and Connell (1979) have concluded that thiamine is destroyed to varying degrees depending upon species and conditions of heat treatment. There is some loss on smoking and complete destruction has been reported during the salting and drying of certain tropical products (Cutting, 1962).

Riboflavin

Riboflavin is heat stable in acid, relatively stable in neutral and is rapidly destroyed in alkaline conditions. It is stable in the presence of oxidising agents but is very sensitive to light (Priestley, 1979). This last point obviously accounts for the observation that riboflavin is completely destroyed during the salting and drying of certain tropical products as well as drying slowly under cool atmospheric conditions (Cutting, 1962). Similarly, Adrian (1957) reported that riboflavin is the only B vitamin which is not essentially retained during the preparation of 14 salted and dried fish products in Angola. Significant concentrations (18 $\mu g/g$), however, were found (Koyama, 1960) in the dark muscle of dried-smoked bonito ('katsuobushi', see Chapter 8).

Pyridoxine

Pyridoxine is stable to heat but is decomposed by light under non-acidic conditions and has been found to be fairly stable in most foods, although data for the destruction of this vitamin are limited and inconsistent. In fact, a report (Schroeder, 1971) which states that losses can be up to 30% in freeze-dried fish has been dubbed suspect (Bluestein and Labuza, 1975).

Vitamin B_{12}

This vitamin is optimally stable at pH 4.5 to 5.0 and is stable to retorting in this region; at pH values above and below it is slightly unstable. It is destroyed by oxidising and reducing agents and by exposure to light; ferrous salts catalyse its destruction while ferric salts stabilise it (Priestley, 1979). Cooking and drying, as in the normal fish meal manufacturing process, does not lead to a reduction of vitamin B_{12} (Lassen, 1965).

Niacin

Fish is a rich source of nicotinic acid. This vitamin is not appreciably affected by heat or light at the pH values normally found in foods; it is arguably the most stable of all the vitamins (Priestley, 1979). Reductions reported during cooking (Aitken and Connell, 1979) can presumably be attributed to leaching or loss in drip.

Pantothenic Acid

This vitamin has a good thermal stability and is optimally stable around neutral pH; as with niacin most reductions found during cooking can be attributed to losses occurring in drip (Priestley, 1979). There is a report (Schroeder, 1971) that freeze-drying fish leads to losses of 20 to 30% but these losses appear to be rather high when compared with other data and are suspect (Bluestein and Labuza, 1975). Excessively heated fish meal has a reduced pantothenic acid content (Mason and Weidner, 1964) but the conditions investigated are most unlikely to occur during normal drying and smoking.

Biotin

Biotin is stable to heat and light under neutral conditions but is unstable at low and high pH. As with the other water soluble vitamins, leaching can account for most of the losses reported during processing of foods and what little evidence there is indicates that it has a good stability (Priestley, 1979).

Folic acid

Free folacin is more stable than the conjugated forms of folic acid; losses during the cooking of foods vary greatly and are due both to thermal destruction and to leaching. The monoglutamate is moderately stable at neutral and in acidic conditions but destruction is accelerated by copper, air and sunlight in the presence of riboflavin (Priestley, 1979). Folic acid in foods is allegedly the B vitamin which is the most seriously affected by heat during cooking; a loss of 46% is reported for halibut (Cheldelin *et al.*, 1943).

CONCLUSION

The evidence cited shows that the different vitamins present in fish are destroyed to varying extents (ranging from slight reductions to almost complete loss) as a result of drying and smoking. These data must be considered, however, in a complete dietary context as was noted at the start of this chapter. The proportion of fish in any person's diet will determine the importance of the contribution made by fish to the total intake of any particular vitamin. Although there are a few exceptions (as with the Japanese and Eskimos) seafoods do not constitute a significant proportion of the human diet and any losses, however large, of vitamins as a result of drying and smoking fish will generally not be of major nutritional importance. Having said that, though, particular care should be taken to ensure that individuals or groups within a population are not at risk because their diets are idiosyncratic.

REFERENCES

Adrian, J. (1957). Composition and nutritional value of fish prepared by various techniques: cured and dried samples from Africa, industrial fish meals and 'nuoc-mam'. *Ann. Nutr. Aliment.* **11**, 27–44.

Aitken, A. and Connell, J.J. (1979). Fish. In: *Effects of Heating on Foodstuffs*, R.J. Priestley (Ed.), Applied Science Publishers, London, pp. 219–54.

Aylward, F. and Haisman, D.R. (1969). Oxidation systems in fruits and vegetables – their relation to the quality of processed products. *Advan. Food Res.* **17**, 1–76.

Bluestein, P.M. and Labuza, T.P. (1975). Effects of moisture removal on nutrients. In: *Nutritional Evaluation of Food Processing*, 2nd Edn, R.S. Harris and E. Karmas (Eds), Avi Publishing Company, Westport, Connecticut, pp. 289–323.

Bramsnaes, F. (1962). The influence of refrigeration and canning on the nutritive value of fish. In: *Fish in Nutrition*, E. Heen and R. Kreuzer (Eds), Fishing News (Books), London, pp. 153–60.

Bunnell, R.H., Keating, J., Quaresimo, A. and Parman, G.K. (1965). Alpha-tocopherol content of foods. *Am. J. Clin. Nutr.* **17**, 1–10.

Castell, C.H., MacLean, J. and Moore, B. (1965). Rancidity in lean fish muscle. IV. Effect of sodium chloride and other salts. *J. Fisheries Res. Board Can.* **22**, 929–44.

Cheldelin, V.H., Woods, A.A. and Williams, R.J. (1943). Losses of B vitamins due to cooking of foods. *J. Nutr.* **26**, 477–85.

Cutting, C.L. (1962). The influence of drying, salting and smoking on the nutritive value of fish. In: *Fish in Nutrition*, E. Heen and R. Kreuzer (Eds), Fishing News (Books), London, pp. 161–79.

Daun, H. (1975). Effects of salting, curing and smoking on nutrients of flesh foods. In: *Nutritional Evaluation of Food Processing*, 2nd Edn, R.S. Harris and E. Karmas (Eds), Avi Publishing Company, Westport, Connecticut, pp. 355–81.

De Ritter, E. (1982). Effect of processing on nutrient content of food: vitamins. In: *Handbook of Nutritive Value of Processed Food*, M. Rechcigl (Ed.), CRC Press, Cleveland, Ohio, vol. 1, pp. 473–510.

Higashi, H. (1962). Relationship between processing techniques and the amount of vitamins and minerals in processed fish. In: *Fish in Nutrition*, E. Heen and R. Kreuzer (Eds), Fishing News (Books), London, pp. 125–31.

Komata, T., Hashimoto, Y. and Mori, T. (1956). B-Vitamins in marine products and their changes in the processing and storage. I. Canned mackerel and tuna in brine. *Bull. Japan. Soc. Sci. Fisheries* **21**, 1236–40.

Komata, T., Hashimoto, Y. and Mori, T. (1957). B-Vitamins in marine products and their changes during the processing and storage. II. Canned baby clam. *Bull. Japan. Soc. Sci. Fisheries* **23**, 44–6.

Koyama, H. (1960). Riboflavin contents in processed protein foods. *J. Japan. Soc. Food Nutr.* **12**, 325–31.

Lassen, S. (1965). Technological problems in the heat treatment of fish requiring more knowledge from fundamental research. In: *The Technology of Fish Utilization*, R. Kreuzer (Ed.), Fishing News (Books), London, pp. 235–40.

Mason, V. C. and Weidner, K. (1964). The effect of heat on the amino acid and B-vitamin composition of fish meal. *Acta Agr. Scand.* **14**, 87–95.

Mejika, A. (1960). Vitamin A of fish meat. *Report Hokkaido Inst. Public Health* **11**, 152–3.

Priestley, R.J. (1979). Vitamins. In: *Effects of Heating on Foodstuffs*, R.J. Priestley (Ed.), Applied Science Publishers, London, pp. 121–56.

Schroeder, H.A. (1971). Losses of vitamins and trace minerals resulting from processing and preservation of foods. *Am. J. Clin. Nutr.* **24**, 562–72.

Tarr, H.L.A. (1969). Nutritional value of fish muscle and problems associated with its preservation. *Can. Inst. Food Technol. J.* **2**, 42–5.

Tilgner, D.J. and Daun, H. (1970). Antioxidative and sensory properties of curing smokes obtained by three basic smoke generation methods. *Lebensm. Wiss. Technol.* **3**(5), 77–82.

Tilgner, D.J., Sikorski, Z.E., Daun, H. and Rudnicki, A. (1965). Antioxidative activity of the dispersed and dispersing phase of curing smoke. XX-Lecie Politech. Gdansk. *Sesja Nauk*. CH45, 344. [Quoted in Daun (1975).]

Van Veen, A.G. (1936). Visch als Voedingsmiddel (Fish as food.) *Geneesk. Tijdschr. Nederl. -Ind.* **76**, 1230–59.

Chapter 5

UNDESIRABLE FACTORS IN DRIED FISH PRODUCTS

B. S. PAN
National Taiwan College of Marine Science and Technology, Taiwan

Even if the desirable flavour, colour and texture of dried fish products are developed through the proper application of heating and/or smoking techniques, there are still undesirable effects which occur simultaneously resulting from the reactions of, and interactions between, the major components of fish flesh. This paper mainly describes the development of certain antinutritional factors in smoked and dried fish, as distinct from reductions in the amounts of nutrients present. Lipid oxidation, protein denaturation and Maillard reactions are the main causes of changes in nutritional quality of dried fish products. Histamine and gizzerosine are the toxic substances that have been found so far in dried fish products specifically in products of migratory fish. *N*-nitrosodimethylamine was found in gas broiled, overheated dried squid but aflatoxin has not been reported as present in dried fish or in fish meals.

DECREASES IN AMINO ACID AVAILABILITY

Nutritional quality may be reduced by the heat treatment which can accompany the drying of fish. It is likely that the availability of amino acids is lowered as a result of their reaction with other components of fish. The Maillard reaction, in which amino compounds react with reducing sugars, occurs during heating and forms a group of uncharacterised brown polymers, melanoidins.

Maillard reactions between glucose and individual amino acids produce fructose–amino acid Amadori compounds. The fructose–methionine Amadori compound was not metabolised by rat as a source of methionine

(Horn *et al.*, 1968). Fructose–tryptophan and fructose–leucine derivatives are absorbed much more slowly than the corresponding free amino acids (Sgarbieri *et al.*, 1973; Tonaka *et al.*, 1975). Fructose–phenylalanine is unavailable to chicks and rats (Johnson *et al.*, 1977, 1979) and the availability of lysine is reduced by forming fructose–lysine (Mori, 1978; Mori *et al.*, 1980). The absorption of fructose–glycine is less than that of free glycine but in the combined state it is still a good non-specific nitrogen source when fed to nitrogen-deprived rats (Hagen *et al.*, 1970).

On the whole, however, fructose–amino acids appear to be metabolised and utilised very little (Erbersdobler *et al.*, 1981; Stegink *et al.*, 1981). Long-term ingestion of a Maillard browned diet produces significant physiological disorders such as vacuolated hepatocytes, accumulations of black-brown pigments in the liver, enlargement and fatty metamorphosis of liver cells (Lee *et al.*, 1981), and alterations in the tubular cells of the kidneys (Erbersdobler *et al.*, 1981).

Anserine, lysine, 1-methyl-histidine and taurine are more reactive with reducing sugars than other amino compounds in cod muscle (Jones, 1962). Among the amino acids which are predominant in squid, lysine and taurine are more susceptible to browning reactions with glucose 6-phosphate than proline, arginine, alanine and glycine (Haard and Arcilla, 1985). Glucose 6-phosphate and ribose 5-phosphate are more reactive in Maillard browning reactions than the corresponding free sugars (Nagayama, 1961; Yamanaka *et al.*, 1974).

Taurine is more than five times as abundant in Atlantic short-finned squid than in other genera of squid, while the latter also exhibit a lower tendency to undergo Maillard browning as a result of heating and dehydration than the former (Haard and Arcilla, 1985).

The losses of available lysine in dried skipjack stick ('katsuobushi'; see Chapter 8) are increased from 16·5% to 30% with increased smoking temperature, from 15 to 90°C (Shiau *et al.*, personal communication).

The availability of lysine can also be lowered by the formation of lysinoalanine during the heating of proteinaceous foods particularly under alkaline conditions. Lysinoalanine has been implicated as causing renal damage in rats although marked differences between species and, within a species, between strains, have been observed; so far it has not been demonstrated that the amounts present in normal diets can affect the health of man.

Lutefisk, a traditional Scandinavian product, is made by treating stockfish (dried cod) with alkali. Lysinoalanine is formed in lutefisk, and in fish protein concentrate, when the pH reaches 13 and is maximised by heating at 90°C (Miller *et al.*, 1983; Fujimaki *et al.*, 1980). Steaming the lutefisk, or prolonged cold storage of the product, causes the lysinoalanine content to increase (Eriksson, 1978). The only other fish product reported to contain lysinoalanine was a sample of canned sardines (Raymond, 1980). It would thus appear that lysinoalanine is not readily formed in fish products but it should be remembered that some of the conditions employed in preparing and cooking dried and smoked fish products could be conducive to the formation of lysinoalanine. For a recent review of this whole subject see Maga (1984).

Prolonged or repeated heating of tuna at temperatures higher than 90°C results in hydrogen sulphide (H_2S) production (Khayat, 1977). The water soluble proteins of tuna contribute 71% of the H_2S while salt soluble proteins generate insignificant amounts (Khayat, 1978). Denaturation is an essential step for the increase in reactivity of the sulphydryl group to produce H_2S (Germs, 1973; Khayat, 1978). The production of H_2S during severe heating is not only significant in flavour deterioration but also indicative of a reduction in cysteine availability.

During storage, dimethyl sulphide (DMS) increases abruptly when the water activity (Aw) of dried laver ('Nori') is above 0·3. No change was observed at Aw below 0·1 (Araki *et al.*, 1983). The increase in DMS is a reflection of a decrease in the content of sulphur amino acids.

DECREASES IN PROTEIN DIGESTIBILITY

The protein digestibility of brown fish meal has been reported as being less than that of white fish meal; in one experiment values of 50% compared with 80 to 85% were obtained (Saruya *et al.*, 1963a). Lipid oxidation is more severe in brown fish meal than in white fish meal. Fish meal prepared by precooking with 0·1% butylated hydroxyanisole (BHA) or by reducing the processing temperature from 90 to 80°C shows improvements in protein digestibility of 4 to 7% (Nose and Toyama, 1966). The peroxide value of the lipids extracted from salted sardine increases during drying and storage (Nakamura *et al.*, 1978). Oxidative rancidity is one of the causes of reduction in protein digestibility for dried fatty fish products such as brown fish meal.

The treatment of brown fish meal with *n*-hexane and a methanol–chloroform mixture results in small increases in protein digestibility and feed efficiency probably due to the removal of some fat soluble negative factors (Nomura *et al.*, 1972).

During storage, the protein digestibility and feed efficiency of brown fish meal are decreased in relation to an increase in oxidised acid content (Saruya *et al.*, 1963*b*). The acid value of brown fish meal is much higher than that of white fish meal (Lain and Chung, 1980; Chung and Lain, 1981). Fish bits (dried shredded fish) also show an increased acid value and reduced lysine availability with development of rancidity (Pan, 1977). It is likely that the acid value may just correlate with the protein digestibility of dried fatty fish products. Lipid oxidation is evidently a cause of the reduced protein digestibility as shown in the changes of fish meal (Saruya *et al.*, 1963*b*).

FORMATION OF MUTAGENS AND CARCINOGENS

The question of carcinogens in smoked fish is covered extensively in Chapter 6 but it is necessary here to include consideration of their occurrence in dried fish products.

Water-soluble, heat-stable, polar substances that are easily washed from fish tissues are responsible for the formation of mutagens during heating (Krone *et al.*, 1986) and purified Maillard reaction products, i.e. reductones, dicarbonyls, pyrazines and furan derivatives, have shown mutagenicity (Powrie *et al.*, 1986).

Mutagens can be formed even by heating fish at 115°C at a relatively high moisture level. The mutagenicity, however, is 2 to 15 times lower than that of the majority of grilled or fried fish and meats (Krone *et al.*, 1986). Pyrolysis of amino acids and proteins in model systems and in grilled meats and grilled fish heated to temperatures above 200°C results in the formation of mutagenic products (Nagao *et al.*, 1977*a*, *b*; Aeschbacker *et al.*, 1981; Yoshida *et al.*, 1986; Terada *et al.*, 1986). During cooking fish or meat, the surface temperature may rise to above 200°C and mutagens such as amino α-carbolines may be formed (Yoshida *et al.*, 1986).

Carcinogenic heterocyclic amines have been identified in broiled sardines and broiled beef (Committee on Diet, Nutrition and Cancer, National Research Council, 1982; Sugimura, 1982; Sugimura and Sato, 1983; Hatch *et al.*, 1984). Low molecular weight (less than 500) compounds in meat are precursors of mutagenic activity in dried meat homogenates oven-baked to 200 or 300°C (Taylor *et al.*, 1986). The consumption of broiled fish containing pyrolysates is significantly and positively associated with deaths from cancer, especially of the stomach (Kuratsune *et al.*, 1986).

FORMATION OF *N*-NITROSODIMETHYLAMINE

N-Nitrosodimethylamine, a carcinogen, is formed when salted dried squid are broiled on a kerosene burner or a gas range. The content increases with heating temperature above 100°C, below which the content is negligible. Broiling on an electric range also produces *N*-nitrosodimethylamine but to a lesser extent (Matsui *et al.*, 1984*a*, *b*).

The development and occurrence of *N*-nitrosodimethylamine and of other *N*-nitrosamines in smoked fish is dealt with extensively by Sikorski (see Chapter 6) but it would be useful at this stage to point out that the risks of formation of compounds of this class would be greatly increased when fish is dried in direct contact with gases of combustion.

MYCOTOXIN FORMATION

Aflatoxin, a carcinogenic mycotoxin, was not detected in 76 samples of fish meal, the water activity of which ranged between 0·41 and 0·51 (B.S. Pan and P.-H. Tung, unpublished data), nor was it found in mouldy dried herring and dried shrimp (B.S. Pan, unpublished data), the water activity of the latter being in the range 0·68 to 0·77 (Pan and Chiang, 1979). Aflatoxins were also not detected by Sim *et al.* (1985) in dried shrimp, some of which had high mould counts and even carried aflatoxin-producing *A. flavus* strains. Other studies, however, have shown that dried fish sampled from tropical areas are highly contaminated by aflatoxin B1 (Okonkwo and Nwokolo, 1978) and other mycotoxins (Enomoto and Saito, 1972). Wrapping in polyethylene, and treatment in 1 % hypochlorite solution reduces the development of aflatoxin in dried fish during storage (Okonkwo and Nwokolo, 1978). Smoking of tilapia also delays and reduces aflatoxin production during storage (Uraih and Ogbadu, 1982). It seems that aflatoxin formation is dependent on the water activity of the dried product, the fungistatic treatment of the product, the type of packaging and storage environment. It is likely that aflatoxin is not produced during the drying process but development during unsatisfactory storage conditions can be a potential hazard.

FORMATION OF HISTAMINE TOXICITY

Toxic levels of histamine have been found in dried and/or smoked products of mackerel, horse mackerel and sardines (Taylor, 1983), as well as in the brown fish meal made from these fish (Toyama *et al.*, 1981; B.S. Pan, unpublished). The high levels of histamine in dried fish products are the result of microbial decarboxylation of the free histidine present in the dark-muscled fish. Exposure of raw fish to high ambient temperatures accelerates the reaction (Pan *et al.*, 1982).

Considerable amounts of the histamine which accumulates in raw fish are pressed out by the 'cook and press' operation in fish meal processing. Fish meal manufactured with addition of fish solubles contains more histamine than dried presscake (Toyama *et al.*, 1982).

Histamine content increases to a maximum then decreases with prolonged drying time (Takagi, 1968). The drying of sardine, previously brined in 5 to 15% NaCl, for 2 h, causes pronounced increases in histamine. Prolonged brining (to 22 h) shows no increase in dried salted sardine (Nakamura *et al.*, 1978). During drying of the presscake to manufacture fish meal, the histamine content decreases due to the Maillard reaction between the amino group of histamine or histidine and reducing sugars. The addition of glucose and ribose reduces the histamine content of the fish meal. Ribose is more favourable to the reaction than glucose (Toyama *et al.*, 1982).

When spoiled tuna containing histamine at a level of 250 mg/100 g is fed to pigs, cats, dogs and rabbits, only Duroc pigs show regurgitation. Those weighing 18 to 36 kg are more sensitive than the 50 to 73 kg pigs (Blonz and Olcott, 1978a). The histamine toxicity in guinea pigs is potentiated by the simultaneous administration of cadaverine (Bjeldanes *et al.*, 1978). *In vitro*, spermine and spermidine inhibit histamine–mucin binding to a greater extent than cadaverine and putrescine. These diamines and polyamines liberate histamine from intestinal mucin and possibly increase histamine absorption *in vivo* (Chu and Bjeldanes, 1982).

Menhaden fish meal fed to rats increases gastric histamine metabolism and acid secretion in the stomach, when compared with diets containing other protein sources (Chen and Fisher, 1983).

The administration of histamine in the diet causes activation of pepsinogen to pepsin, increased stomach acidity and increased cathepsin D activities in the spleen and muscle indicative of a muscle softening mechanism and stomach erosion or ulceration in mackerel (Otake *et al.*,

1977; Otake, 1979), an abnormality in the stomach of rainbow trout and gizzard erosion in broiler chickens (Watanabe *et al.*, 1982).

In addition, uptake of histamine accelerates the idioventricular rhythm and sinus rate in guinea pig heart (Levi, 1979). Since human heart has histamine receptors (Ginsburg, 1980; Bristow, 1982), it becomes a target organ in systemic allergic reactions (Capurro, 1975).

Weight gains are depressed when quails are fed with dried spoiled tuna (Blonz and Olcott, 1978*b*). Preliminary tests showed that tilapia fed for a week with diets containing 2000 ppm histamine lose appetite and become very slow in moving (B.S. Pan, unpublished).

Aspects of the occurrence of histamine in spoiling fish are discussed in detail by Pan and James (1985).

INDUCTION OF GIZZEROSINE

Feeding rainbow trout with brown fish meal heated at 130°C for 3 to 5 h results in stomach erosion (Watanabe *et al.*, 1982). The causative substance, named gizzerosine, has recently been identified as 2-amino-9-(4-imidazolyl)-7 azanonanoic acid, a toxic derivative of histamine or histidine; it is responsible for producing gizzard erosion (GE) and 'black vomit' in chicks (Okazaki *et al.*, 1983).

Fish meal prepared with putrid mackerel and sardine promotes GE more readily than that prepared from fresh fish. The dry rendering of fish scrap releases more lipid and results in more severe lipid oxidation in fish meal and worse GE in the chicks fed than the wet rendering of round fish. The addition of the antioxidant, EMQ, to fish prevents the formation, during heating and drying, of 'GE causative substances', while the presence of histamine in raw material accelerates their formation (Toyama *et al.*, 1982).

REFERENCES

Aeschbacker, H.U., Chappius, C.H., Manganel, M. and Aesback, R. (1981). Investigation of Maillard products in bacterial mutagenicity test systems. In: *Maillard Reactions in Foods*, C. Eriksson, (Ed.), Pergamon Press, Oxford, pp. 279–94.

Araki, S., Ogawa, H., Oohusa, T., Kayama, M. and Kobayashi, A. (1983). Flavor of dried laver 'Nori' at different water activities, with special reference to dimethyl sulfide production. *Bull. Japan Soc. Sci. Fisheries* **49**(11), 1717–20.

Bjeldanes, L. F., Schutz, D.E. and Morris, M.M. (1978). On the aetiology of scombroid poisoning: cadaverine potentiation of histamine toxicity in the guinea-pig. *Food Cosmet. Toxicol.* **16**, 157–9.

Blonz, E.R. and Olcott, H.S. (1978*a*). Effects of orally ingested histamine and/or commercially canned spoiled skipjack tuna on pigs, cats, dogs and rabbits. *Comp. Biochem. Physiol.* **61**C, 161–3.

Blonz, E.R. and Olcott, H.S. (1978*b*). Canned spoiled tuna on growth in young Japanese quail. *J. Food Sci.* **43**(5), 1390–1.

Bristow, M.R. (1982). Histamine and the human heart: the other receptor system. *Am. J. Cardiol.* **49**, 249–51.

Capurro, N. (1975). The heart as a target organ in systemic allergic reactions; comparison of cardiac anaphylaxis *in vivo* and *in vitro*. *Circulation Res.* **36**, 520–8.

Chen, S.C. and Fisher, H. (1983). Gastric histamine metabolism and acid secretion in rats as influenced by diet and nutrient content. *J. Nutr.* **113**, 379–87.

Chu, C.H. and Bjeldanes, L.F. (1982). Effect of diamines, polyamines and tuna fish extracts on the binding of histamine to mucin *in vitro*. *J. Food Sci.* **47**, 79–88.

Chung, C.Y. and Lain, J.L. (1981). Comparative studies on the quality of brown and white fish meals II. Quality evaluation of commercial eel feeds sold in Taiwan. *Nat. Sci. Counc. Monthly, ROC* **9**(4), 354–64.

Committee on Diet, Nutrition, and Cancer, National Research Council (1982). Mutagens in food. In: *Diet, Nutrition, and Cancer*, Chapter 13, National Academy Press, Washington DC, pp. 1–27.

Enomoto, M. and Saito, M. (1972). Carcinogens produced by fungi. *Ann. Rev. Microbiol.* **26**, 279–312.

Erbersdobler, H.F., Brandt, A., Scharrer, E. and von Wangenheim, B. (1981). Transport and metabolism studies with fructose amino acids. In: *Maillard Reactions in Foods*, C. Eriksson (Ed.), Pergamon Press, Oxford, pp. 257–64.

Eriksson, C.E. (1978). Protein recovery and lysinoalanine formation in Scandinavian traditional, alkaline fish production. In: *Proceedings of the Fifth International Congress on Food Science and Technology*, Abstract, p. 72.

Fujimaki, M., Haraguchi, T., Abe, K., Homma, S. and Arai, S. (1980). Specific conditions that maximize formation of lysinoalanine in wheat gluten and fish protein concentrate. *Agr. Biol. Chem.* Tokyo **44**(8), 1911–6.

Germs, A.C. (1973). Hydrogen sulfide production in eggs and egg products as a result of heating. *J. Sci. Food Agr.* **24**, 7–16.

Ginsburg, R. (1980). Histamine receptors in the human heart. *Life Sci.* **26**, 2245–9.

Haard, N.F. and Arcilla, R. (1985). Precursors of Maillard browning in Atlantic short finned squid. *Can. Inst. Food Sci. Technol. J.* **18**, 326–31.

Hagen, S.N., Horn, M.J., Lipton, S.H. and Womack, M. (1970). Fructose-glycine as a source of non specific nitrogen for rats. *J. Agr. Food Chem.* **18**, 273–5.

Hamm, R. and Hofmann, K. (1965). Changes in the sulfhydryl and disulfide groups in beef muscle proteins during heating. *Nature* **207**, 1269.

Hatch, F.T., Felton, J.S., Stuermer, D.H. and Bjeldanes, L.F. (1984). Identification of mutagens from the cooking of food. In: *Chemical Mutagens: Principles and Methods for their Detection*, Vol. 9, F.J. de Serres, and A. Hollaender, (Eds), Plenum Publishing, New York, pp. 111–64.

Horn, J.M., Linchtenstein, H. and Womack, M. (1968). A methionine-fructose compound and its availability to microorganisms and rats. *J. Agr. Food Chem.* **16**, 741–5.

Johnson, G.H., Baker, D.H. and Perkins, E.G. (1977). Nutritional implications of the Maillard reaction: the availability of fructose-phenylalanine to the chick. *J. Nutr.* **107**, 1659–64.

Johnson, G.H., Baker, D.H. and Perkins, E.G. (1979). Nutritional implications of the Maillard reaction. The metabolism of fructose-phenylalanine in the rat. *J. Nutr.* **109**, 590–6.

Jones, N.R. (1962). Browning reactions in dried fish products. In: *Recent Advances in Food Science, Vol. 2, Processing*, J. Hawthorn and J.M. Leitch (Eds), Butterworths, London, pp. 74–80.

Khayat, A. (1977). Hydrogen sulfide production by heating tuna meat. *J. Food Sci.* **42**(3), 601–9.

Khayat, A. (1978). Hydrogen sulfide production by heating different protein fractions of tuna meat. *J. Food Biochem.* **2**, 121–31.

Krone, C.A., Yeh, S.M.J. and Iwaoka, W.T. (1986). Mutagen formation during commercial processing of foods. *Environmental Health Perspectives* **67**, 75–88.

Kuratsune, M., Ikeda, M. and Hayashi, T. (1986). Epidemiologic studies on possible health effects of intake of pyrolyzates of foods, with reference to mortality among Japanese Seventh-Day Adventists. *Environmental Health Perspectives* **67**, 143–6.

Lain, J.L. and Chung, C.Y. (1980). Comparative studies on the quality of brown and white fish meals. I. Studies on the quality of imported brown fish meal and its comparison with imported white fish meal. *Nat. Sci. Counc. Monthly, ROC.* **8**(4), 346–54.

Lee, T.-C., Kimiagar, M., Pintauro, S.J. and Chichester, C.O. (1981). Physiological and safety aspects of Maillard browning of foods. In: *Maillard Reactions in Foods*, C. Eriksson (Ed.), Pergamon Press, Oxford, pp. 243–256.

Levi, R. (1979). Acceleration of idioventricular rhythm by histamine in guinea pig heart. Mediation by H_2 receptors. *Circulation Res.* **44**, 847–55.

Maga, J.A. (1984). Lysinoalanine in foods. *J. Agr. Food Chem.* **32**(5), 955–64.

Matsui, M., Ishibashi, T. and Kawabata, T. (1984*a*). Effect of broiling temperatures on the formation of *N*-nitrosodimethylamine from dried squid products. *Bull. Japan. Soc. Sci. Fisheries* **50**(1), 151–4.

Matsui, M., Ishibashi, T. and Kawabata, T. (1984*b*). Precursors of N-nitrosodimethylamine formed in dried squid upon broiling. *Bull. Japan. Soc. Sci. Fisheries* **50**(1), 155–9.

Miller, R., Spinelli, J., Babbitt, J.K. (1983). Effect of heat and alkali on lysinoalanine formation in fish muscle. *J. Food Sci.* **48**(1), 296–7.

Mori, B. (1978). ϵ-Fructoselysine as an absorption-delayed material in the small intestinal lumen of rats fed browned casein. *Agr. Biol. Chem* (Tokyo) **42**, 2071–6.

Mori, B., Kojima, K. and Saito, S. (1980). ϵ-Fructoselysine in urine of rats fed ^{14}C-lysine-labelled casein browned by amino-carbonyl reaction. *Agr. Biol. Chem.* (Tokyo) **44**, 1327–31.

Nagao, M., Honda, M., Seino, Y., Yahagi, T. and Sugimura, T. (1977*a*).

Mutagenicities of smoke condensates and the charred surface of fish and meat. *Cancer Lett.* **2**, 221–6.

Nagao, M., Yahagi, T., Kawachi, T., Seino, Y., Honda, M., Matsukura, N., Sugimura, T., Wakabayashi, K., Tsuji, K. and Kosuge, T. (1977*b*). In: *Progress in Genetic Toxicology*, Vol. 2, D. Scott, B.A. Bridges, and F.H. Sobels, (Eds), Elsevier/North-Holland, Amsterdam, pp. 259–264.

Nagayama, F. (1961). Studies on the browning of fish flesh-IV. Browning of sugar with certain amino acids by heat. *Bull. Japan. Soc. Sci. Fisheries* **27**, 28–33.

Nakamura, K., Fujii, Y. and Ishikawa, S. (1978). Studies on salted and dried sardine-I. Changes of the chemical components in sardine meat during salting, drying and storage. *Bull. Tokai Reg. Fisheries Res. Lab.* **95**, 75–83.

Nomura, M., Nose, T., Uematsu, Z. and Inaba, D. (1972). Application of imported Peru fish meal in fish feed-I. Feeding experiment with rainbow trout. *Bull. Freshwater Fisheries Res. Lab.* **22**(1), 85–113.

Nose, T. and Toyama, K. (1966). Protein digestibility of brown fish meal in rainbow trout. *Bull. Freshwater Fisheries Res. Lab.* **15**(2), 213–24.

Okazaki, T., Noguchi, T., Igarashi, K., Sakagami, Y., Seto, H., Mori, K., Naito, H., Masumura, T. and Sugahara, M. (1983). Gizzerosine, a new toxic substance in fish meal, causes severe gizzard erosion in chicks. *Agr. Biol. Chem.* **47**, 2949–52.

Okonkwo, P.O. and Nwokolo, C. (1978). Aflatoxin B1: simple procedures to reduce levels in tropical foods. *Nutr. Reports Intern.* **17**(3), 387–95.

Okuzumi, M., Awano, M. and Ohki, Y. (1984). Effects of temperature, pH value and NaCl concentration on histamine formation of N-group bacteria (psychrophilic and halophilic histamine-forming bacteria). *Bull. Japan Soc. Sci. Fisheries* **50**(10), 1757–62.

Otake, S. (1979). Mechanism of the biochemical changes in mackerel muscle deterioration. *New Food Industry* **24**(10), 54–61.

Otake, S., Maeda, T. and Fukui, K. (1977). Biological investigation on the metabolism of histamine in mackerel (*Scomber japonicus*) – I. Influence of histamine injection into stomach on the breakdown of the tissues. *Bull. Japan Soc. Sci. Fisheries* **43**(4), 477–88.

Pan, B.S. (1977). Low-moisture fishery products. In: *Fishery Products of Taiwan*, J.L. Chuang, B.S. Pan and G.C. Chen (Eds), JCRR Fisheries Series 25B, Taipei, Taiwan, pp. 43–54.

Pan, B.S. and Chiang, W.-C. (1979). Quality evaluation and changes during storage of commercial dried salted shrimps. *Food Sci.* (Chinese) **6**(1), 24–35.

Pan, B.S. and James, D. (Eds) (1985). *Histamine in Marine Products: Production of Bacteria, Measurement and Prediction of Formation.* FAO Fisheries Technical Paper 252, FAO, Rome, 62 pp.

Pan, B.S., Yang, B.H. and Huang, D.M. (1982). Histamine formation in canning of bonito and mackerel. *Food Industry* **14**(2), 84.

Powrie, W.D., Wu, C.H. and Molund, V.P. (1986). Browning reaction systems as sources of mutagens and antimutagens. *Environmental Health Perspectives* **67**, 47–54.

Raymond, M.L. (1980). Studies concerning the determination of lysinoalanine in food proteins. *J. Food Sci.* **45**, 56–9.

Saruya, K., Inaba, D., Toyama, K., Nomura, M., Yamamoto, M. and Koyama,

S. (1963*a*). Feed efficiency of fish meal – I. Relation between oxidation and feed efficiency of fish meal. *Bull. Japan Soc. Sci. Fisheries* **29**(10), 948–52.

Saruya, K., Inaba, D., Arai, G., Toyama, K., Nagashima, Y. and Koyama, S. (1963*b*). Feed efficiency of fish meal – II. Digestibility of fish meal treated with various solvents. *Bull. Japan Soc. Sci. Fisheries* **29**(10), 953–6.

Sgarbieri, V.C., Amaya, J., Tonaka, M. and Chichester, C.O. (1973). Nutritional consequences of the Maillard reaction. Amino acid availability from fructose-leucine and fructose-tryptophan in the rat. *J. Nutr.* **103**, 657–63.

Sim, T.S., Teo, T. and Sim, T.F. (1985). A note on the screening of dried shrimps, shrimp paste and raw groundnut kernels for aflatoxin-producing *Aspergillus flavus*. *J. Appl. Bact.* **59**(1), 29–34.

Stegink, L.D., Freeman, J.B., Denbesten, L. and Filer, L.J. Jr. (1981). Maillard reaction products in parenteral nutrition. In: *Maillard Reactions in Foods*, C. Eriksson (Ed.), Pergamon Press, Oxford, pp. 265–78.

Sugimura, T. (1982). Mutagens in cooked food. *Basic Life Sci.* **21**, 243–269.

Sugimura, T. and Sato, S. (1983). Mutagens-carcinogens in foods. *Cancer Res.* (Suppl.) **43**, 2415s–21s.

Takagi, M. (1968). Fundamental studies on food poisoning caused by sea foods I. On the fate of histamine contents during the drying of fish meat. *Bull. Fac. Fisheries*, Hokkaido Univ. **18**, 385–91.

Taylor, S.L. (1983). *Monograph on Histamine Poisoning*. Codex Alimentarius Commission, FAO, Rome.

Taylor, R.T., Fultz, E. and Knize, M. (1986). Mutagen formation in a model beef supernatant fraction. IV. Properties of the system. *Environmental Health Perspectives* **67**, 59–74.

Terada, M., Nagao, M., Nakayasu, M., Sakamoto, H., Nakasato, F. and Sugimura, T. (1986). Mutagenic activities of heterocyclic amines in Chinese hamster lung cells in culture. *Environmental Health Perspectives* **67**, 117–20.

Tonaka, M., Lee, T. and Chichester, C.O. (1975). Nutritional consequences on the Maillard reaction. The absorption of fructose-L-tryptophan in the large intestine of the rat. *J. Nutr.* **105**, 989–94.

Toyama, K., Okuzumi, M., Yokoi, T. and Aoe, H. (1981). Histamine content of fish meal. *Bull. Japan Soc. Sci. Fisheries* **47**(3), 415–9.

Toyama, K., Hoshi, M., Ishiguro, M. and Aoe, H. (1982). Behavior of histamine during the manufacturing process of fish meal. *Bull. Japan Soc. Sci. Fisheries* **48**(9), 1333–9.

Uraih, N. and Ogbadu, L. (1982). Influence of woodsmoke on aflatoxin production by *Aspergillus flavus*. *Eur. J. Appl. Microbiol. Biotechnol.* **14**(1), 51–3.

Watanabe, T., Takeuchi, T., Satoh, S., Niimi, S., Toyama, K. and Okuzuni, M. (1982). Effect of histamine and its derivative on rainbow trout. Presentation at the Annual Meeting of Japanese Society of Scientific Fisheries, April 1982.

Yamanaka, H., Bito, M. and Yokoseki, M. (1974). Orange discolored meat of canned skipjack – III. Changes in amounts of glycolytic intermediates in skipjack meat during thawing and cooking. *Bull. Japan, Soc. Sci. Fisheries* **40**, 941–7.

Yoshida, D., Matsumoto, T., Okamoto, H., Mizusaki, S., Kushi, A. and Fukuhara, Y. (1986). Formation of mutagens by heating foods and model systems. *Environmental Health Perspectives* **67**, 55–8.

Chapter 6

SMOKING OF FISH AND CARCINOGENS

Z. E. Sikorski
Technical University, Gdansk, Poland

INTRODUCTION

In the last three decades much attention has been paid to the hazards associated with the presence of carcinogenic or mutagenic compounds in foods that have been charcoal-broiled, grilled, baked in gas ovens, or smoked. The conditions prevailing during the roasting, broiling and smoking of meats and fish may be favourable for generating carcinogenic polycyclic aromatic hydrocarbons (PAH), *N*-nitroso compounds (NNC), and heterocyclic aromatic amines. Most of the available published information regarding these compounds in fish deals with the presence of PAH while the NNC have been mainly investigated in cured and smoked meats. The heterocyclic aromatic amines are either derived from amino acid and protein pyrolysates or are Maillard reaction products, and are found mainly in grilled meats and fish and are not associated with the smoking process *per se*. (See, however, Chapter 5: section on mutagens and carcinogens.)

POLYCYCLIC AROMATIC HYDROCARBONS

The Source of PAH in Smoked Fish

Many PAH have been found in smoked meats and fish. Not all of them are carcinogenic but the compound benzo(a)pyrene (BP) is regarded as an indicator of carcinogenicity. The presence of this compound has been thoroughly investigated in many foods and it is also limited by the regulations of some countries.

Various fresh seafoods contain detectable amounts of PAH. In vertebrate fish no PAH could be detected (Dunn and Fee, 1979) even in samples from polluted fishing areas. This was related by the authors to the ability of fish to metabolise PAH and excrete them as water soluble derivatives. In commercial samples of crab and shrimp Dunn and Fee found up to 0·5 ng/g BP and in lobsters 0·8–7·9 ng/g. Lobsters which had been kept in a commercial tidal pond made of creosoted timber, however, contained much higher amounts of chrysene, benzo(a)-anthracene, benzo(b)fluoranthene, dibenz(ah)anthracene and indeno-(1,2,3-cd)pyrene reaching, in the case of BP, 2300 ng/g wet weight in the digestive gland and 281 ng/g in the edible tail meat.

In processed fish the contents of PAH may be higher, as these compounds originate from various sources. Canned mackerel in vegetable oil was found to contain up to 2·6 ng BP per gram wet weight (Dunn and Fee, 1979) and the oil was suspected to be the source of the PAH.

In smoked fish the PAH originate from wood smoke. The results of extensive investigations showed that the hydrocarbon fraction of wood smoke contains a large variety of PAH. According to Tilgner and Daun (1969) different investigators have identified 24 PAH and about 40 unidentified compounds were detected. Ten years later Potthast separated the components of the isocyclic hydrocarbon fraction of wood smoke by capillary gas chromatography and identified 30 compounds by mass spectrometry (Potthast, 1979).

It has been known from ages-old trade as well as from industrial experience that the composition and suitability of smoke for treating different foods depends on the variety and properties of the wood and on the conditions of smoke generation, i.e. the temperature of combustion, access of air and purification of the smoke. The influence of these parameters on the formation of different smoke components was investigated by Pettet and Lane (1940), Spanyar *et al.* (1960), Jahnsen (1961), Rusz (1962), Tilgner and Miler (1963), Tilgner and Daun (1965), Kurko (1969), Borys (1977), Obiedziński (1977), Olkiewicz (1977), Potthast (1978) and several other workers. The aim of these investigations was to find conditions for producing smoke with the highest concentration of desirable phenols and carbonyl compounds but containing the lowest amounts of PAH.

The influence of temperature on the formation of PAH in smoke was observed in early investigations (Miler, 1958, 1962). By using a two-stage generator, in which the thermal decomposition of wood and the oxidation of the volatile decomposition products were carried out in separate reactors, it was established that at high temperatures PAH were formed from both lignin and cellulose. To prevent the formation of BP it was found necessary to keep the temperatures of decomposition and oxidation below 400 and 200°C, respectively (Tilgner and Miler, 1963; Tilgner and Daun, 1965). In later experiments carried out in Miler's group (Borys, 1977; Obiedziński, 1977; Olkiewicz, 1977) it was shown that the rate of smouldering of sawdust was a linear function of the product of temperature and air velocity in the smoke generator. At constant temperature the smouldering rate is a function of air flow and can be represented by a Gaussian type of curve. The temperature of the smouldering bed depends on the air flow rate. Thus it is possible to keep industrial smoke generators at a desired constant temperature by precise control of the air flow. The rates of formation of phenols, carbonyl compounds, and PAH are functions of the product of temperature and air velocity and can also be represented by Gaussian-type curves. The peaks of the curves for these groups of compounds do not coincide and their locations along the temperature–air flow velocity axis depend also on the humidity of the sawdust and the kind of wood. It is thus possible to find, for any generator, specific operating conditions at which the formation of PAH is effectively suppressed, while that of phenols and carbonyl compounds is favoured.

The Contents of PAH in Smoked Fish

Much information on the presence of different PAH and on the contents of BP in smoked fish has been published. According to data compiled by Tilgner and Daun (1969) the concentration of BP in smoked fish ranged from 0·7 to 60 ng/g wet weight. In the edible parts of hot smoked Baltic cod and sprat from traditional kilns Wierzchowski and Gajewska (1972) found 0·5–5·3 and 1–3·5 ng BP per gram, respectively.

The content of PAH in the edible parts of smoked fish depends obviously on the concentration of these compounds in the wood smoke and on the parameters of smoking, as well as on the species and preparation of the fish. The content of BP is higher in fish smoked in traditional kilns than in fish produced in smokehouses supplied with smoke from separate generators working at low temperatures. Petrun and Rubenchik (1966) found 0·7–1·7 ng/g BP in electrostatically smoked sprats and from 4·15 to 60 ng/g in hot and cold smoked fish (cod, horse mackerel, mackerel, humpback salmon) from traditional kilns. The smoke for electrostatic smoking was generated at 275–300°C. Benzo(a)pyrene was determined in the mince of whole sprats, including skin, while in the other, larger fish only the edible parts were analysed. It is evident from

TABLE 1

THE CONTENT OF BENZO(A)PYRENE (BP) IN SMOKED FISH (STEINIG, 1976)

Product	*BP (ng/g) in fish smoked in*			
	Traditional kilns		*Smokehouses supplied from external smoke source*	
	Edible parts	*Skin*	*Edible parts*	*Skin*
Hot smoked fish				
Sprat	1·1–2·4	9–28	—	—
Buckling	0·47–1·4	22–43	—	—
Mackerel	0·53–2·4	19–30	0·47–0·94	1·3–2·4
Eel	2·6–3·3	49–74	0·33–0·53	1·8–4·0
Halibut without skin	1·5–3·7	—	0·64–0·76	
Dogfish without skin				
total edible part	2·6–3·7	—	1·1	—
outer layer	15–16	—	5·0–8·9	—
inner layer	0·88–2·0	—	0·47–0·53	—
Cold smoked fish				
Salmon without skin				
outer layer	3·6	—	0·77–4·7	—
inner layer	0·62	—	0·24–0·67	—

these results, as well as from data presented by Steinig (1976) (see Table 1), that the influence of the parameters of smoke generation and of smoking on the BP content in the product is very distinct. The role of the skin as a barrier for PAH is also well illustrated in Table 1.

In market samples of canned smoked sprats the content of BP in the meaty part was found to be 0·64–7·3 ng/g and in the oil 4·2–21 ng/g, while in canned mackerel and salmon it was less than 1·6 and 4·7 ng/g, respectively (Steinig, 1976).

Detailed published data on the concentrations of other PAH in smoked fish are not available. It may, however, be assumed that the quantitative relationship between different PAH in smoked fish should be similar to that in smoked meat products. The investigations of Potthast (1979) revealed that smoked meat products contain a large number of low molecular weight PAH, mol. wt below 216, which are less harmful than PAH with higher molecular weights. The low molecular weight PAH are

TABLE 2
THE CONTENTS OF BENZO(A)PYRENE (BP) IN SELECTED SMOKED MEAT PRODUCTS (POTTHAST, 1979)

Meat product	*Number of samples*	*Content of BP ng/g*	*Proportion of samples containing BP > 1ppb %*
Black smoked cooked ham	60	0·10–15·60	60
Black smoked uncooked ham	28	0·10–15	25
Dry sausage in natural casing	53	0·10–5·42	2·5
Hot smoked and scalded sausage	36	0·10–2·08	5·5

generally present in the smoked meat products at much higher concentrations, up to 50 times higher than that of BP, whilst the concentrations of the more carcinogenic higher molecular weight compounds are similar to that of BP. The total mass of the carcinogenic high molecular weight PAH is, on average, 10 times that of BP. According to Potthast (1979) the average contents of BP in different smoked meat products are usually below the limit of 1 ng/g set in the Federal Republic of Germany (Table 2).

A comparison of the available data indicates that the contents of PAH in the edible parts of fish smoked in modern smokehouses is not higher than that in smoked meat products. Discussing the possible health hazards caused by smoked fish it is necessary to consider also other dietary sources of PAH and the level of consumption of smoked fish, that is the contribution of smoked fish to the estimated daily dietary intake of PAH. It is also necessary to take into consideration the presence of other substances modifying the carcinogenic activity of PAH.

The contents of BP in many other foods is higher than in some smoked fish, e.g. baked bread and biscuits, normal roasted coffee, soya bean oil, and barbecued ribs may contain respectively 0·5, 0·3–0·5, 3·5, and 10·5 ng BP in 1 gram (Gangolli, 1986; Steinig, 1976). To exert their carcinogenic effect PAH require metabolic activation. It is known that vitamin A and antioxidants like BHA and BHT inhibit some PAH-induced carcinogenicity (Gangolli, 1986). Fatty fish are rich sources of vitamin A and wood smoke contains many very potent phenolic antioxidants (Sikorski, 1980). According to Gangolli (1986) 'Consideration of the relative minor contribution of PAHs derived from smoked foods on the estimated daily dietary intake *per capita* in the UK, and of data on the biological effects of orally ingested PAHs in experimental animals, together with the known protective effects of components in the human diet, leads one to a conclusion that PAHs in smoked foods are unlikely to present a major carcinogenic hazard to man'.

N-NITROSO COMPOUNDS

The Sources of NNC in Fish

Some representatives of this group of potent carcinogens have been known since the 1960s to be present in fishery products. At that time hepatotoxic disorders were found in Norway in farm animals fed fish meals produced from fish offal that had been preserved with nitrite.

The carcinogenic *N*-nitrosamines (NNA) are not indigenous in fish tissues. They are formed by the reaction of nitrogen oxides with, mainly, secondary amines present in the fish. The nitrogen oxides are generated from nitrites by the following reactions; the relative amounts and reactivity of each species being dependent upon acidity.

$$2HO{-}N{=}O \leftrightharpoons O{=}N{-}O{-}N{=}O + H_2O$$

$$O{=}N{-}O{-}N{=}O + H^+ \leftrightharpoons O{=}N{-}\underset{\displaystyle H}{\underset{|}{\overset{+}{O}}}{-}N{=}O \rightleftharpoons \overset{+}{N}{=}O + HO{-}N{=}O$$

Nitrogen oxides are also present in wood smoke. The *N*-nitrosation proceeds as follows:

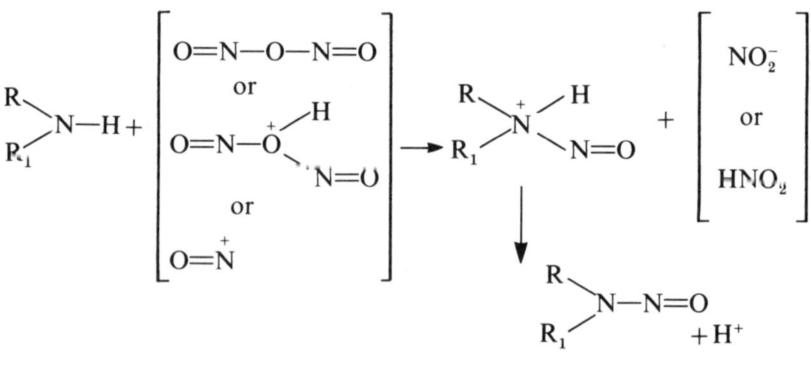

TABLE 3

THE CONTENT OF *N*-NITROSAMINES (NNA) IN SMOKED FISH PRODUCTS (DIKUN *et al.*, 1980)

Fish commodity	*Content of NNA (ng/g)*		
	Total NNA	*NDMA*	*NDEA*
Baltic sprat			
frozen	6·6–20·5	20·3	1·8
hot smoked	5·6–32	105·0	10·0
canned in oil			
meaty part	5·0–31·0	—	—
oil	29·0–86·0	—	—
total contents of can	10·2–27·4	41·6	4·1
Herring			
frozen	16·5	12·5	1·4
hot smoked	22·0	5·7–9·4	0·6–0·8
canned in oil			
meaty part	15·5–28·0	1·5–2·6	0
oil	47·5–103·3	0·3–0·5	0
total contents of can	27·5–44·8	1·2–2·3	0
Cod			
frozen	38·8	21·6	4·2
hot smoked	136	67·1	1·0
canned in oil			
meaty part	40·5	12·3	0
oil	71·0	40·0	0
total contents of can	45·2	16·5	0

Fish products are known to be rich in amines but rather poor sources of nitrites. Sodium nitrite has been generally prohibited as a preservative in the fish industry and nitrates are used only to a limited extent in a few countries in preparing some products, e.g. anchovies or other pickled commodities. According to Nieper and Etzel (1983) rainbow trout, fed on feed mixtures containing about 500 ppm sodium nitrate, accumulated from 17 to 229 ppm sodium nitrate in the muscle. In commercial smoked fish, however, the major source of nitrogen oxides is the wood smoke.

The Contents of NNC in Smoked Fish

Several different fish products have been reported to contain detectable quantities of NNA (Iyengar *et al.*, 1975). Because of the use of nitrates in

TABLE 4

THE CONTENT OF *N*-NITROSODIMETHYLAMINE (NDMA) IN SMOKED FISH (RÖPER *et al.*, 1981)

Fish	*Smoking*	*NDMA ng/g*
Halibut	Hot	1·3
Herring bloater	Hot	1·4
Mackerel	Hot	8·6
Eel	Hot	55·8
Salmon	Cold	Not detectable

TABLE 5

THE CONTENT OF VARIOUS *N*-NITROSO COMPOUNDS IN FOODS

Food commodity	*Compound ng/g*	*Reference*
	N-nitrosodimethylamine	
Meats, smoked	2–35	Ravesi (1976)
Cod fillets, raw	0·2–0·4	Key *et al.* (1982)
Fish, cooked in gas oven	0·1–0·9	Key *et al.* (1982)
Malt, dried in gas drier at 120°C	80–320	Fritz and Uhde (1982)
Beer, Pils	0·5–7·0	
	N-nitrosopyrolidine	
Meats, smoked	13–105	Ravesi (1976)
Bacon	10–108	Ravesi (1976)
Bacon fat, cooked out	45–207	Ravesi (1976)
	N-nitrosothiazolidine	
Bacon, raw	N–7·5	Sen *et al.* (1985)
Bacon, fried	N–241	Sen *et al.* (1985)
Bacon fat, cooked out	N–2·5	Sen *et al.* (1985)
	N-nitrosothiazolidine-4-carboxylic acid	
Bacon, raw	64–9000	Sen *et al.* (1985)
Bacon, fried	20–14 000	Sen *et al.* (1985)
	Total *N*-nitroso compounds as N NO	
Mutton, smoked, cured, and cooked	2500–17 700	Dennis *et al.* (1984)

N = negative, < 1ppb.

the curing of titbits and anchovies it would be expected that these products would contain detectable amounts of NNA. Pedersen and Meyland (1981) found *N*-nitrosodimethylamine (NDMA) in commercial samples of various Scandinavian pickled fish at concentrations up to 2·2 ng/g. Iyengar *et al.* (1975) and Röper *et al.* (1981), however, did not detect NNA in pickled herring or anchovy, respectively.

Many data on the contents of NNA in smoked fish were presented by Dikun *et al.* (1980) (Table 3). The total NNA was determined colorimetrically, while NDMA and *N*-nitrosodiethylamine (NDEA) were determined by gas liquid chromatography. The identification of the chromatographic peaks was not confirmed by mass spectrometry. No other available published results are so high, especially as regards unsmoked frozen fish. The data published by Röper *et al.* (1981) are significantly lower, except for the high value for hot smoked eel (Table 4).

Deliberating on the possible health hazards induced by NNC present in smoked fish one should again consider the aspects presented above in the case of PAH. The concentration of NNA in smoked fish is not significantly higher than in other foods (Table 5). Furthermore, some NNC can be degraded during sterilisation (Dikun *et al.*, 1980) or metabolised by some microorganisms (Harada and Yamada, 1979) thus reducing the amounts likely to be ingested.

REFERENCES

Borys, A. (1977). Wpływ warunków zżarzania trocin na zawartość związków karbonylowych w dymie. (The effect of sawdust pyrolysis on the contents of carbonyl compounds in smoke.) Thesis, Instytut Przemysłu Mięsnego i Tłuszczowego, Warszawa.

Dennis, M.J., Cripps, G.S., Tricker, A.R., Massey, R.C. and McWeeney, D.J. (1984). *N*-Nitroso compounds and polycyclic aromatic hydrocarbons in Icelandic smoked cured mutton. *Food Chem. Toxicol.* **22**(4), 305–6.

Dikun, P.P., Romanova, L.E., Sheidrikova, J.A., Reim, I.D. and Yunosova, Z. Yu. (1980). Soderzhanie N-nitrosoaminov v nekoterykh vidakh rybnykh konservov. (The contents of *N*-nitrosamines in some varieties of canned fish.) *Rybnoe Khozyaistvo* **8**, 69–73.

Dunn, B.P. and Fee, J. (1979). Polycyclic aromatic hydrocarbon carcinogens in commercial seafoods. *J. Fisheries Res. Board Can.* **36**, 1469–76.

Fritz, W. and Uhde, W.J. (1982). Zur Bestimmung von Nitrosaminen im Malz und Bier. (On the determination of nitrosamines in malt and beer). *Lebensmittel industrie* **29**, 209–15.

Gangolli, S.D. (1986). The toxicology of smoked foods. *IFST Proc.* **19**(2), 68–78.

Harada, K. and Yamada, K. (1979). Microbial degradation of nitrosamines. I. *Bull. Japan Soc. Sci. Fisheries* **45**, 925–8.

Iyengar, J.R., Panalaks, T., Miles, W.F. and Sen, N.P. (1975). A survey of fish products for volatile *N*-nitrosamines. *J. Sci. Food Agr.* **27**, 527–30.

Jahnsen, V.J. (1960). The chemical composition of hardwood smoke. Thesis, Purdue University.

Key, P.E., Bayly, J.M., Massey, R.C. and McWeeny, D.J. (1982). Nitrosodimethylamine levels in fish cooked by natural gas and by electricity. *J. Food Technol.* **17**, 703–8.

Kurko, V.I. (1969). *Khimiya kopcheniya.* (The chemistry of smoking.) Pishcheprom, Moscow.

Miler, K. (1958). Hypothetical possibilities of removal of cocarcinogens from curing smoke. First Session on Curing Smoke. Politechnika Gdańska, Gdańsk.

Miler, K. (1962). Możliwosci wytwarzania dymu wędzarniczego wolnego od 3,4-benzopirenu i 1,2,5,6,-dwubenzoantracenu. (The possibilities of making curing smoke containing no 3,4-benzpyrene and 1,2,5,6-dibenzanthracene.) Thesis, Politechnika Gdańska, Gdańsk.

Nieper, L. and Etzel, V. (1983). Zum Nitratgehalt in der Muskulatur von Räucherfischen. (On the contents of nitrate in the muscles of smoked fish.) *Archiv fur Lebensmittelhygiene* **34**, 149–50.

Obiedziński, M. (1977). Wpływ warunków zżarzania drewna na zawartość benzo(a)pirenu w dymie drzewnym. (The effect of wood pyrolysis on the contents of benzo(a)pyrene in wood smoke.) Thesis, Instytut Przemysłu Mięsnego i Tłuszczowego, Warszawa.

Olkiewicz, M.J. (1977). Wpływ warunkow zżarzania drewna na zawartość frakcji fenolowej w dymie wędzarniczym. (The effect of wood pyrolysis on the contents of the phenolic fraction in wood smoke.) Thesis, Instytut Przemysłu Mięsnego i Tłuszczowego, Warszawa.

Pedersen, E. and Meyland, J. (1981). Nitrate, nitrite, and volatile nitrosamine in pickled fish prepared with addition of nitrate. *Z. Lebensm. -Untersuch. -Forsch.* **173**, 359–61.

Petrun, A.S. and Rubenchik, B.L. (1966). O vozmozhnosti vozniknoveniya kantserogennogo veshchestva 3,4-benzpirena v rybe, podvergnutoi elektrokopcheniyu. (On the possibility of appearance of the carcinogenic compound 3,4-benzopyrene in electrostatically smoked fish.) *Vrachebnoe Delo* **5**, 93–5.

Pettet, E.E.J. and Lane, F.G. (1940). A study on the chemical composition of wood smoke. *J. Soc. Chem. Ind.* **59**, 114–19.

Potthast, K. (1978). Verfahren des Räucherns und ihr Einfluss auf den Gehalt an 3,4-Benzpyren und anderen Inhaltsstoffen des Räucherrauches in Geräucherten Fleischerzeugnissen. (The procedure of smoking and its effect on the contents of 3,4 benzpyrene and other smoke constitutents in smoked meat products.) *Fleischwirtschaft* **58**, 340, 347–8.

Potthast, K. (1979). Einflus der Räuchertechnologie auf die vollständige Zusammensetzung der polycyclischen Kohlenwasserstoffe in geräucherten Fleischwaren, in Rauchkondensaten und in den Abgasen von Räucheranlagen. (The influence of smoking technology on the composition of polycyclic hydrocarbons in smoked meat products, smoke condensates and in waste gases from smoking plants.) *Fleischwirtschaft* **59**, 1515–23.

Ravesi, E.M. (1976). Nitrite additives – harmful or necessary? *Marine Fisheries Rev.* **38**(4), 24–30.

Röper, H., Heyns, K. and Gunther, W. (1981). Nitrosamine trace analysis of anchovies and smoked fish. *Chem. Mikrobiol. Technol. Lebensm.* **7**, 13–17.

Rusz, J. (1962). Investigation of suitable conditions in the production of smoke for the smoking of meat products. *Technologia Mesa.* (Spec. ed.), 10–15.

Sen, N.P., Seaman, S.W. and Baddoo, P.A. (1985). *N*-nitrosothiazolidine and non-volatile *N*-nitroso compounds in foods. *Food Technol.* **39**(1), 84–8.

Sikorski, Z.E. (1980). Technologia żywności pochodzenia morskiego. (Marine food technology.) Wyd. Naukowo Techniczne, Warszawa.

Spanyar, P., Kevei, E. and Kiszel, M. (1960). Über Fragen des Räucherns von Lebensmitteln. II. Zusammensetzung des Rauches und dessen Beeinflussung durch Faktoren der Rauchbildung. (About problems of smoking of foods. II. Composition of smoke and the effect of parameters of smoke generation.) *Z. Lebensm. -Untersuch. -Forsch.* **112**, 471–80.

Steinig, J. (1976). 3-4-Benzpyren-Gehalte in geräucherten Fischen in Abhängigheit von der Räuchermethode. (The contents of 3,4-benzopyrene in smoked fish as influenced by the smoking method.) *Z. Lebensm. -Untersuch. -Forsch.* **162**, 235–42.

Tilgner, D.J. and Daun, H. (1965). Wpływ celulozy i ligniny na obecność składników rakotwórczych w aerozolu wędzarniczym. (The effect of cellulose and lignin on the presence of cancerogenic components in the smoking aerosol.) *Roczniki Państwowego Zakładu Hig.* **16**, 45–52.

Tilgner, D.J. and Daun, H. (1969). Polycyclic aromatic hydrocarbons (polynuclears) in smoked foods. *Residue Rev.* **27**, 19–41.

Tilgner, D.J. and Miler, K. (1963). Możliwości eliminacji związków rakotwórczych z dymu wędzarniczego. (Possibilities of eliminating carcinogens from smoke used in the manufacture of smoked food.) *Przemysł Spożywczy* **17**, 85–90.

Toth, L. (1982). *Chemie der Räucherung.* (The Chemistry of Smoking.) Verlag Chemie, Weinheim.

Wierzchowski, J. and Gajewska, R. (1972). Oznaczanie 3,4-benzopirenu w rybach wędzonych. (Determination of 3,4-benzopyrene in smoked fish.) *Bromatol. Chem. Toksykol.* **5**, 483–6.

Chapter 7

PROCESSING AND STORAGE OF TRADITIONAL DRIED AND SMOKED FISH PRODUCTS

R. G. Poulter
Overseas Development Natural Resources Institute, London, UK

INTRODUCTION

The following section is intended as a brief guide to traditional processing methods. Details of the processing techniques vary from region to region and sometimes even from processor to processor. The objective of processing is to provide some degree of preservation and also to give a product the desired eating characteristics. The degree of preservation required can vary considerably and this determines the degree of processing or curing which is considered desirable. Hence a distinction may be made between the so-called hard cured products which are well preserved, and the soft cured products which have relatively short shelf lives unless supplementary preservation methods such as chilling or freezing are also used. For hard cured products preservation is the primary objective of processing while for soft cured products improving the eating qualities of the products is as significant as the preservation effect.

In many developing countries fish processed by traditional methods are extremely popular and continue to be a major source of inexpensive dietary protein. This popularity is in part due to the chemical changes which occur during the smoking and drying processes; the chemicals produced being, in fact, responsible for the products' characteristic (and desired) flavours. The preferences that people have can be traced to familiarity with traditional products such as those produced by heavy smoking over open fires or by drying extensively in direct sunlight. In many instances where the traditional process may have been superseded and nutritionally less degraded products have become available, the preference for the flavours of the traditional products remains in the population, and there is a risk that attempts to attain a product that would again satisfy such a preference could very likely be at the expense of the nutritional advantages recently gained.

On the other hand, some taste preferences have changed over the years and hard cured products have become less popular in most developed countries. The modern mildly smoked and less extensively dried products are less stable than the traditional ones and this trend has been made possible by the introduction of alternative preservation methods such as chilling and freezing. Also, as the drying process becomes increasingly sophisticated, and hence more expensive, the operators tend to reduce drying times and at the same time improve product yield by retaining more moisture in the product; but this would be at the expense of keeping quality if unfrozen.

A list of some general references to accounts of processing and storing traditional dried and smoked fish products is given at the end of this chapter.

DRESSING

Some preparation before salting, drying and/or smoking is generally performed; splitting and gutting are conventional treatments. Only small fish would be processed without any dressing at all. For heavily salted fish scoring the flesh may also be necessary while for some soft cured products fillets may be used.

SALTING

Fish can be salted in three basic ways, as follows.

1. Kench curing, where solid salt is rubbed into the fish flesh, the fish then being stacked while the salt penetrates the flesh and the extracted moisture drains away.

2. Pickling, which is similar to kench curing except that the moisture extracted from the fish is not allowed to drain. The fish becomes immersed in a very salty 'pickle' of extracted fluids.

3. Brining, where fish are soaked in a prepared concentrated salt solution.

The length of salting can vary considerably from just a few minutes to several weeks or even months.

SUN AND AIR DRYING

Fish can be dried after salting by any of the above methods, or it can be dried unsalted. Small fish or split large fish are simply spread in the sun in tropical countries allowing moisture to evaporate from the flesh surface. Fish may be laid directly on the ground, on mats or on raised racks. In colder climates fish can be hung up to dry in the wind, this process being slower than sun drying but the lower ambient temperatures slow the spoilage processes.

Turning the fish from time to time during drying aids the process by exposing more of the surface of the fish to the air. Too rapid an initial drying results in the surface flesh forming a hard crust (case hardening) which slows the rate of diffusion of water to the outer surface of the fish. Bacteria can multiply in the moist interior which has a high water activity, causing the centre to spoil while the outside appears satisfactory. Drying times vary considerably depending on the process and weather; 3 to 10 days are generally required to prepare a typical sun-dried product. The final moisture content of the dried product can also vary considerably depending on the desired storage and eating qualities of the product.

SMOKING

There are basically three types of smoked fish products:

1. cold smoked in which the fish is not cooked;
2. hot smoked in which the fish is cooked; and
3. smoke-dried in which the fish is cooked and subsequently dried.

Fish are sometimes pre-treated before smoking starts. European smoked products are often brined and sometimes slightly dried before smoking; traditional African smoked products are rarely brined, but some may be sun-dried before smoking starts.

Cold smoked fish are processed at low temperatures (30 to 40°C depending on species) to prevent cooking or coagulation of protein. The process is designed to give the desired flavour to the product rather than a significant degree of preservation. Products of this type are not traditional in tropical countries. Hot smoked products are cooked, a process which denatures enzymes, practically sterilises the product and hence imparts some degree of preservation. Products of this type are found both in tropical countries and temperate countries. They can be consumed without any further cooking by the consumer although they may be cooked again prior to consumption. Neither cold nor hot smoked fish will keep for long periods unless supplementary techniques, such as drying, freezing and chilling are used.

Smoke-dried products are popular in many parts of the developing world, most especially in Africa. Processing starts with hot smoking but instead of finishing the process with a moist product it is further smoked to remove moisture. The final product is dried sufficiently to ensure an extended storage life. Drying is achieved by an extended smoking time or by a series of short smoking periods over several days.

STORAGE AND DISTRIBUTION

During the distribution of cured fish the products may have to be stored for some time. This can occur where there is a need to collect large quantities of fish together for the next step in the distribution chain or as part of marketing strategy. During times of glut traders will not sell their products as they will fetch very low prices so they will store them until fish is in short supply and prices are favourable. Lengths of storage can therefore, vary enormously. Some cured products are consumed within days of processing while for others there may be a delay of several months, possibly a year, before they are consumed. In some situations it may be known in advance that the products will have to be stored for a considerable period but in other situations the storage time is unpredictable due to marketing factors. It is quite common for products to be processed initially for a short storage period and then have to be reprocessed as the storage period becomes extended, for example by re-drying or re-smoking.

Storage facilities vary tremendously. In some areas permanent structures with weather-proof roofs and concrete floors may be found, while in other areas fish may be stacked up on the ground, uncovered, or piled up on drying racks. It may be packed in simple packaging materials such as gunny sacks, palm leaves or bamboo cartons, or it may be stacked up loose in storage areas and packaged only prior to onward distribution.

By the time the fish arrives at the market it may have become mixed with products from many processing sites. The fish may be sorted into product types such as smoked and unsmoked, salted and unsalted and a very limited attempt at sorting by quality may be practised by the individual trader. At the market stall it is generally kept loose in baskets or on mats so that purchasers can select pieces by hand.

CONCLUSIONS

Fish curing is an important industry in many tropical developing countries even though it has declined in some areas as a result of the increased use of more sophisticated preservation techniques. The processes are characteristically labour intensive but can be performed with limited facilities making them especially appropriate for fisheries which lack the necessary infrastructure to use other techniques. Traditionally cured fish is, and will continue to be, a major source of dietary protein in many developing countries where malnutrition is, unfortunately, all too common.

BIBLIOGRAPHY

Brownell, B. and Lopez, J. (1985). The Chorker fish smoking method: a truly appropriate method. In: *FAO Expert Consultation on Fish Technology in Africa*, F11: FTA/85/20, FAO, Rome, 11 pp.

Burgess, G.H.O. and Bannerman, A.McK. (1963). *Fish Smoking: a Torry Kiln Operator's Handbook*, HMSO, London, 44 pp.

Clucas, I.J. (1981). *Fish Handling Preservation and Processing in the Tropics: part 1*, Report of the Tropical Development and Research Institute G144, London, 141 pp.

Clucas, I.J. (1982). *Fish Handling Preservation and Processing in the Tropics: part 2*, Report of the Tropical Development and Research Institute G145, London, 145 pp.

Curran, C.A. (1984) *Storage of Cured Fishery Products in Malaysia*, Report of the Tropical Development and Research Institute L71, London, 67 pp.

Cutting, C.L. (1955). *Fish Saving: A History of Fish Processing from Ancient to Modern Times*, Leonard Hill, London, 372 pp.

Diouf, N. (1980). *Trials on Protection of Dried Fish by Pyrethrum and Solar Energy*, FAO conference, Dar-es-Salaam, Tanzania, FAO, Rome, 22 pp.

Doe, P.E., Hashmi, R., Poulter, R.G. and Olley, J. (1982). Isohalic sorption isotherms I. Determination for dried salted cod (*Gadus morrhua*). *J. Food Technol.* **17**, 125–34.

Doe, P. E., Curran, C.A. and Poulter, R.G. (1983). Determination of the water activity and shelf life of dried fish products. In: *The Production and Storage of Dried Tropical Fish*, D. James (Ed.), FAO Fisheries Report No. 279 Suppl., FAO, Rome, pp. 201–21.

Trim, D.S. and Curran, C.A. (1983). *A Comparative Study of Solar and Sun Drying of Fish in Ecuador*, Report of the Tropical Development and Research Institute L60, London, 44 pp.

Wood, C.D. (1981). *The Prevention of Losses in Cured Fish*, FAO Fisheries Technical Paper No. 219, FAO, Rome.

Wood, C.D. (1985*a*). Methodology for the assessment of losses of cured fish and the evaluation of countermeasures. In: *FAO Expert Consultation on Fish Technology in Africa, Lusaka, Zambia, Jan 1985*, Tropical Development and Research Institute, London.

Wood, C.D. (1985*b*). Losses of traditionally cured fish in Africa. In: *FAO Expert Consultation on Fish Technology in Africa, Lusaka, Zambia, Jan 1985*, Tropical Development and Research Institute, London.

Wood, C.D., Curran, C.A., Poulter, R.G. and Doe, P.E. (1982). Reduction of losses in cured fish products through improved control of the process. In: *FAO Expert Consultation on Fish Technology in Africa, Lusaka, Zambia, Jan 1985*, Tropical Development and Research Institute, London.

Chapter 8

EFFECT OF SMOKING AND DRYING ON THE NUTRITIVE VALUE OF FISH: A REVIEW OF JAPANESE STUDIES

T. MOTOHIRO
Kagoshima University, Kagoshima, Japan

SMOKED FISH

The European method of smoking was initially imported for salmon into Japan in 1871–2. Since then, smoked salmon or herring have been manufactured and become increasingly popular. In addition, smoked squid meat has been increasingly consumed.

The smoked fish produced in this country can be put into two categories: cold and hot. The process of cold smoking was originally developed to introduce a characteristic preservative function by wood smoke to the fish rather than to modify food quality such as flavour, taste and texture. On the other hand, hot smoking was developed to give smoke flavour to the fish, with an aim of producing particular sea foods with improved food quality but not to give preservability. Hot smoking is further divided into three categories: mild or high temperatures and short time processes. The following are the fish species which are used in smoked sea food produced in Japan.

Cold smoking: herring, salmon, yellowtail, cod and mackerel.
Hot smoking:
Mild temperature process: salmon, eel, cod, oysters, herring, Alaska pollack and Atka mackerel.
High temperature process: squid and octopus.
Short time process: salmon and rainbow trout.

Generally each process consists of several steps. The fish is prepared, salted, desalted, washed and drained, dried, smoked, and finalised.

Cold Smoking

In general, gutted and eviscerated fish are dry salted to make the flesh harden; they are then desalted to a level of salinity which does not give too salty a taste. The fish are transferred to the kiln and hung over a fire of smouldering sawdust. The distance between the nearest fish and the fire should be determined by the water and the salt contents of the fish. Cold smoking is carried out at 15°C to 30°C (average 25°C) for 4–6 weeks. During the smoke-drying process, the rates of cooling down and drying of the fish are essentially controlled during the day by introducing warm and humid air to the kiln. The fish is exposed to the smoke generated at the furnace by closing the ventilator attached to the kiln at night. Cold smoked fish is seldom produced in the summer season since the temperature required is lower than 25°C.

Ignition is made at three points in the furnace for the first week, after which smouldering starts, at four points for the second week, and five to six points for the third and later weeks. Then two pieces of firewood and 7·5 kg of sawdust are put onto the bottom of the furnace and ignition is made by burnt charcoal. The shutter of the ventilator put in the ceiling is fully opened for the earlier stages of smoke-drying, and is then closed to some extent during the middle stages, and is eventually closed almost completely for the period from the third week to the end. On rainy days, however, the shutter is kept continuously closed to prevent the fish from reabsorbing water. The temperature should be kept at 18°C for the first week in the kiln, 22°C for the second week and 25°C for the final step.

Hot Smoking

There are two different ways of preparing fish for hot smoking. In the first, if the fish have already been salted, they are desalted: in the second, if they are fresh or thawed, they have to be salted. The salt content must be lower than for cold smoking.

Generally, if the fish is as large as salmon, fillets with skin are prepared by removing head, tail, viscera and bones. If it is of a much larger size, the fish body is cut into small blocks and if it is of a smaller size, whole body (so-called bloater) or eviscerated and split body (so-called kipper fillets) products are prepared.

The whole fish or the fillets are usually subjected to the smoking process immediately after splitting. There is another procedure, however, in which they are dipped in a solution containing salt, flavouring materials and spices and are then subjected to the smoking process.

The hot smoking process is performed within the temperature range of 30–80°C. If the smoking process is within 30–50°C, it is called mild temperature smoking; the smoking process carried out at 50–80°C is called high temperature smoking.

The time required for hot smoking is much shorter than that for cold smoking, and is complete within 3–8 h. Occasionally, the smoking and drying processes are continued alternately for 2–3 days by lowering temperature to some extent in order to give preservability to the final product. The fish subjected to the smoking process is allowed to stand in the smoking chamber so that the temperature reduces gradually.

It is essential that the temperature in the kiln is controlled carefully for hot smoking and, to achieve this, handling fish and operating the kiln are usually performed during the day.

A comparatively large quantity of sawdust is piled onto firewood to cause smouldering. Seven to ten fires are put in one furnace, and approximately seven pieces of firewood are placed on one fire. The fish hanging in the furnace are placed as near as possible to the fire and care taken to avoid broiling.

High temperatures, such as 50–80°C, are applied occasionally during the processing of either hot smoked animal or boiled fish meat. The exposure to the smoke is usually continued until a smoke flavour is given to the fish. The temperature is maintained by either electric or steam heaters which are thermostatically controlled. The humidity of the warm smoke can be controlled by altering the amount of fresh air entering the kiln.

The warm smoke is blown by a fan at an even speed over trolleys of fish in a horizontal tunnel. A proportion of it then passes up a chimney but most is circulated again. Half way through the process, the trolleys of fish are interchanged because those nearest the incoming stream of smoke dry more rapidly than the rest.

Although a great deal of smoked fish is still produced by means of traditional kilns, mechanical kilns are being rapidly adopted in Japan. Kilns of various capacities and types are available. Use of a mechanical kiln, however, does not guarantee satisfactory results.

In the following sections the various smoked and dried fishery products to be found in Japan are described. The methods used to prepare them are summarised but additional information can be found in Tanikawa (1985).

Cold Smoked Salmon

The salmon used are mostly King salmon (*Oncorhynchus tschawytscha*) and red salmon (*O. nerka*). Occasionally, chum salmon (*O. keta*) are used.

The salmon are gutted, eviscerated, washed and dry-salted. The quantity of salt and the time of curing vary with the size of the fish. The curing is performed generally by using solid salt at the rate of 20–30% of the weight of fish, and by leaving them for 10–30 days. On some occasions they are cured in saturated brine.

After curing, the fish are desalted in a wooden barrel for 1–3 days with running or repeatedly changed water. The desalted fish are hung onto tenters with the heads down and then dried in the shade until the surface skin has dried.

The dried fish are then smoked in a smokehouse during the night. During the day, the door of the smokehouse is opened for cooling and for drying the fish. For the early stages of smouldering, the duct shutter attached to the ceiling is opened to assist the drying of the fish.

When the fish are half-dried, the shutter is opened slightly; it is eventually closed almost completely to give a smoky flavour to the fish. The temperature inside the smokehouse is about 18°C for the first week, about 22°C for the second week, and about 25°C at the final stage.

For cold smoked red salmon the body is cut to remove head and belly portions.

Cold Smoked Yellowtail

Cultured or wild yellowtail is used to prepare the cold smoked product. The fish are filleted and, if they are not cultured fish, a fillet is separated into back and belly portions. After sufficient bleeding, the bones in the belly portion are removed and parts of belly and the dark meats are trimmed away. The trimmed fillets are cured in brine containing 25 to 30% of salt, 5% of sugar and 2% of spice to the weight of the meat; they are placed in a container and pressed with a weight. Curing is continued for 8–15 days for the wild yellowtail, and for 4–5 days for the cultured fish. During the curing process, the position of the fillets is changed at least once. After curing, the fillets are desalted in fresh water for 2–3 days and for 1 day for wild and cultured yellowtail, respectively. The water is changed repeatedly during desalting.

The desalted fillets are hung on tenters in such a way that the head part hangs down: they are dried in the smokehouse for 1 day. For the wild yellowtail, the shutter in the ceiling is opened fully for the first 4–5 days at 20–25°C to dry the fillets, it is half-closed for the next 3 days at 25° to 30°C to smoke-dry the fillets, and is then closed for the final 3 days at 30–35°C to smoke the fillets. Drying for 1 day is carried out at each step throughout the smoking process. With cultured yellowtail, the smoking time is shortened by a few days.

Cold Smoked Herring

Round herring is usually used for this smoked product. Since the viscera are perishable, however, gutted and eviscerated fish are occasionally used. For dry-salting, fish are placed in a barrel into which salt, equivalent to 12–20% of the weight of the fish, is added, the top of the piled fish is covered and a weight is added. The weight is increased gradually and curing allowed to proceed for a week. For brining, fish are dipped in saturated brine into which additional salt, equivalent to about 5% of the weight of the fish, is supplied. A light weight is applied and curing takes about 40 h. The cured fish are desalted in fresh water which is repeatedly changed. That part of the head near the eyes is pierced with a thin iron bar and the fish are arranged along the bar. Then the bars with the arranged fish are hung onto tenters. The fish are dried and smoked according to a procedure similar to that used for smoked salmon. The smoking process is completed after 3–4 weeks. The yield is about 45%.

Hot Smoked Salmon

Frozen red salmon is usually used. The half-thawed fish are beheaded, eviscerated and filleted. The fillets are cured in a 15–18% brine for several hours, desalted, dried and smoked. The smoking process depends upon the size and freshness of the fish; the initial stage lasts for 7 h at 50°C, and then continues for 2–3 h at 80°C.

Hot Smoked Herring

The herring are cut at the back, eviscerated, washed with dilute brine, and cured in 10–15% brine for 40–90 min. The brined fish are drained, dried and smoked. Smoking time is usually 4–6 h, and the temperature is about 20°C at the initial stage and is then increased gradually to reach 85°C at the final stage.

Seasoned-Smoked Squid or Octopus Meat

As a method of utilising squid meat, in addition to dried squid ('surume'), seasoned-smoked squid ('ika-kunsei') is prepared abundantly in Japan, especially in Hakodate. Octopus or cod meat is also used for preparing such products. The processing of the seasoned-smoked squid meat is as follows:

Due to a decrease of squid caught in the waters around the coast of Japan, squid processors use 'surume-ika' (*Todarodes pacificus*) and 'aka-ika' (or 'baka-ika' and 'murasaki-ika') (*Ommastrephes bartrami*) as well as imported squids such as 'New Zealand surume-ika' (*Nototodarus sloani sloani*), 'Tasmanian surume-ika' (or 'Australian surume-ika') (*Nototodarus sloani gouldi*), 'Canadian irekkus' (or 'matsu-ika', 'Canadian matsu-ika', and 'Newfoundland ika') (*Illex illecebrosus*), 'toli-ika' (*Synplectoteuthis onalaniensis*) and 'tsume-ika' (*Onychoteuthis boreali japonicus*).

Fresh or frozen squid are used. The cephalopodium and visceral mass are removed from the body of the squid. At this stage the tendons are also removed. The fin is cut off at the point of attachment. The removed parts are used for preparation of fish manure. The dressed bodies are thoroughly washed with fresh water.

To accomplish skinning, the dressed bodies are soaked in hot water at 50–55°C, being agitated so that they rub against each other. In the hot water the epidermis is dissolved by enzymatic action. It requires about 10–20 min of soaking to skin them completely. Stale squid require only about 10 min for skinning. The skinned squid are washed and boiled at 80–90°C for 2–3 min.

The boiled squid are seasoned by sprinkling with seasoning materials. The proportion of the seasoning materials is as follows for 3·75 kg of boiled squid: salt 200–250 g, sugar 750–940 g, sodium glutamate 37 g and sodium 5′-ribonucleotide 3·7 g. The bodies sprinkled with the seasoning material are piled in two or three layers on a table, and light pressure is applied using stones in order to cause the seasoning material to penetrate rapidly into the meat. The seasoned squid are hung from wooden bars in a smokehouse.

Smoking is carried out at 20–25°C for the initial 1–2 h, at 50–60°C for the next 4–6 h, and at 60–70°C for the final 2–3 h. It requires about 7–9 h to obtain good smoking. The smoked squid are cut into small rings of 1–2 mm width.

The squid meat is seasoned by the second seasoning. At this time a seasoning solution comprised of 375 g of sugar, 100–110 g of salt, 3·7 g of sodium glutamate, 0·4 g sodium 5′-ribonucleotide and 500 ml of water is sprinkled upon the meat. The meat is then put in a stainless steel box and agitated in order to distribute the seasoning materials. If the surface of the seasoned pieces of meat are wet, they are dried over a charcoal fire or by infrared rays.

Each 1870 g of the finished products are packaged in plastic film bags and sterilised in a steaming box at 85–90°C for 30–60 min.

DRIED FISH

The dried products are prepared for the purpose of preservation, but more recently they have also been used as raw material for seasoned foods.

Dried marine products may be classified as follows:

1. Plain dried products: fish or algae are dried in the sun or in a dryer after being carefully washed. These products include squid ('surume'), octopus ('hoshi-dako'), shark fin ('fuka-bire'), air bladder ('gyoto'), cod ('botara'), herring ('migaki-nishin'), pressed sardine ('tatami-iwashi'), tetraodon-fish (globe fish; 'hoshi-fugu'), herring roe ('kazunoko'), laminaria ('konbu'), laver ('hoshi-nore'), undaria ('wakame') and 'funori'.
2. Boiled and dried products: fish are dried after being boiled. These products include abalone (amber or dark grey colour) ('kai-ho'), sea cucumber (with or without papillae) ('iriko'), sardine ('niboshi-iwashi'), shark cartilage ('meikotsu'), shrimp ('hoshi-ebi'), shark fin ('taishi'), adductor muscle of shellfish ('hoshi-kaibashira'), sand lance ('niboshi-ikanago') and shellfish meat ('niboshi-kai').
3. Broiled and dried fish: fish that are dried after being broiled, include sea-bream ('hamayaki-dai'), 'ayu' ('yakiboshi-ayu') and sardine ('yaki-boshi-iwashi').
4. Dried-salted fish: fish is salted slightly and dried under the sun or in a dryer. Such products include sardine ('shio-boshi-iwashi'), cod ('shio-boshi-tara'), yellowtails ('waramaki-buri'), pompanos (jack mackerel) ('enkan-aji') and mullet roe ('karasumi').
5. Frozen and dried products: fish or the gelatinous extract of algae is frozen and thawed and dried several times. Such products are cod or Alaska pollack ('mintai') and agar-agar ('kantan').
6. Boiled, roasted and dried fish ('fushi'): fish are boiled after beheading and eviscerating, and are then roasted in an oven and dried. Bonito (*Katsuwonus pelamis*), tuna (*Thynnus sibi*) and mackerel are processed in this manner.
7. Seasoned and dried products: fish fillets or sliced meat of molluscs is

seasoned with the seasoning juice (soya bean sauce ('shoyu'), sugar, sweet 'sake', etc.), and then dried by smoking or in a dryer. Such products are relatively new compared with smoked squid meat or filleted Alaska pollack meat.

Plain Dried Products (Air-drying)

Dried Squid

Large quantities of squid or cuttle fish are dried annually as 'surume'. There are many important species caught in Japan.

The process for preparing 'surume' consists of cutting, washing, drying and shaping. The belly side of the mantle-cartilage of the squid is split out from the neighbourhood of the infundibulum to a point distant from the end of the mantle-cartilage by means of a knife with the edge held upward. Then the body, including the cephalopodium, becomes a flat triangular-shape. The ink-sac attached to the visceral mass is removed by hand. Then the visceral mass (of which the greater part is liver) is cut off from the muscular part of the cephalopodium with a knife and discarded. The head is split at the central line to remove the jaws of the mouth. At this time the transparent tendonous matter (*Sepia esculenta* has a large hard shell) is removed with a knife. This should preferably be done before 10.00 am in order that the material may be partially dried the first day.

The cut and split squid are soaked in fresh water for a short time to whiten the surface of the body. They are then washed with sea water or water containing salt (Bé 2° to 3°) to remove mucous and other substances. They are finally washed with fresh water to remove salt in order to prevent absorption of water after drying.

Drying is generally carried out in the sun. It is completed in 3 days when the weather is favourable. The squid hang on straw ropes which are stretched between the upright wooden poles of a wood bar frame construction. The washed squid are placed on the straw ropes in such a manner that the fin of the body is on one side and the tentacles on the other. Drying is carried on from about 10.00 am to 4.00 pm. If the weather is fine, at about 3.00 pm two squid adjoining each other are tied together using a tentacle of each. Each pair of squid are transferred from the straw rope to rafters (about 4–6 cm diameter; 80 squids per rafter), which are brought into a hut and are covered with a straw mat before sunset. If the weather is fine the next morning, the rafters on which the half-dried squid hang are brought out of doors and the drying is continued. When they are two-thirds dried, the squid are taken from the rafter and each is shaped into the desired appearance.

Groups of six shaped squid tied together by their tentacles, are again hung on the rafters. On the third day, if the weather is fine, the drying is complete. In the afternoon, each squid is shaped again and groups of 10 squid each are tied together by their tentacles. The tied squid are packed into a straw bag, and exported after inspection and grading by a Government Inspector.

Dried Herring (*'Migaki-Nishin'*)

Two methods of preparation are employed: the back meat fillet method and the side meat method.

a. Back meat fillet method. Twenty herring from which gill, roe, milt and other visceral masses have been removed, are tied together by a straw rope passed through the gill slits to the mouths; they are then hung over a wooden bar and dried for 2–3 days. Then the back bone is removed from the caudal fin to the head. Next, the belly flesh is cut off horizontally along the lower line of the back bone. The dressed fish is then dried for 2–3 weeks by air-drying. Thus, three parts, namely back flesh, belly flesh and back bone, are obtained. The back flesh is called 'migaki-nishin' (cut-off meat).

b. The side meat method. The herring taken from the wooden bars after 2 days' drying are filleted on both sides of the back bone. Thus two side fillets (left and right) are obtained from a herring body.

Dried Sardines

There are many kinds of *Clupeidae* and *Engraulidae* in Japan. Of the sardines (*Clupeidae*), there are: herring ('nishin', *Clupea pallasi*), sardine ('ma-iwashi', *Sardinia melanosticta*), round herring ('urume-iwashi', *Etrumeus micropus*), *Stelephorus japonicus* ('kibinago'), *Sardinella zunasi* ('sappa') and *Ilisha elongata* ('hira'). To the *Engraulidae* belong the *E. japonicus* ('seguro-iwashi') and anchovy ('katakuchi-iwashi', *Anchovia indicus*). Among the sardines, sardine ('ma-iwashi'), round herring ('urume-iwashi') and anchovy ('katakuchi-iwashi') are air-dried in Japan. The small anchovy is dried and is regarded as a palatable dish. Boiled-dried sardines are also prepared for consumption by the domestic market.

The dried anchovy ('tazukuri') is processed as follows: small anchovies, 6–9 cm in length, are washed in fresh water. After washing the fish are spread on straw mats for wind-drying, being turned over several times in a day. At night the rolled straw blinds in which the anchovies are held are brought indoors. The drying may be finished in several days. The yield of dried anchovy is about 20–30% of the weight of the raw material. The small sardine ('ma-iwashi') is also dried similarly to the dried anchovy.

Dried Cod and Alaska Pollack

Cod and Alaska pollack are wind and air-dried making dried round cod, cod stomach, Alaska pollack and pollack stomach. The dried round Alaska pollack is called 'mintai' in Korea, where it is dried by repeated freezing during winter nights and thawing under winter day light. This is classified as a type of frozen-dried product.

a. Dried round cod ('*bo-tara*'). Cod are beheaded, eviscerated and the back bone is removed; the fish are then dried in the sun until they become as stiff as sticks. They should be dried to a moisture content of less than 30%.

b. Dried round Alaska pollack. This product is prepared in the same manner as cod.

Boiled-Dried Products

Raw marine products (animal or plant) are boiled in order to destroy their enzymatic activity and to remove fatty substances; they are then dried by natural or artificial drying methods. Processes used in preparation of boiled-dried products are described below in the order of their volume of production.

Boiled-Dried Sardines ('*Niboshi-Iwashi*')

Sardine ('ma-iwashi', *Sardinia melanosticta*) and anchovy ('katakuchi-iwashi', *Anchovia indicus*) are used for boiled-dried products as well as for canning. Boiled-dried sardines are used mainly for preparing soup stock ('dashi').

Sardines or anchovies about 10 cm in length are washed in water to remove the scales and impurities. The washed fish are placed in a row on a 'seiro' (steaming basket) of which the bottom is made from bamboo. A stack of 13–17 baskets is placed in boiling salt water in a large iron pot by means of a hoist. The salt water is made by adding 1–1·2 kg of salt to 20 litres of water. Boiling is continued until the fish float (about 15 min). After boiling, the fish are drained by tipping the baskets. They are then completely dried on straw mats for 2–3 days. The dried sardines are packaged in paper bags or boxes according to size and quality.

Boiled-Dried Sand Lance (*'Niboshi-Ikanago'*)
Sand lance ('ikanago', *Ammodytes personatus*) has many different names according to the place of production. The sand lance is a small fish having a length of 15 cm, and lives along almost all shores of the Japanese Islands. This fish is used mainly for boiled-dried products. 'Tsukuda-ni' (seasoned fish) is prepared from the boiled-dried sand lance.

The processing of the boiled-dried sand lance is almost the same as that for boiled-dried sardine. The yield is 15–20% of the weight of the raw fish.

Broiled-Dried Products
Broiled-dried fish are prepared, in general, for the purpose of short time preservation. Broiled-dried fish may consist of sea-bream, flat fish, sardine and other fish.

The process of preparing broiled-dried sea-bream is as follows: the scales on the surface of the fish are first removed and then the visceral mass is removed through the gill slit. The eviscerated fish is then washed with fresh water. One end of a bamboo stick is pierced from mouth to gill slit. The opposite end of the bamboo stick is stuck into the ash in a fire box and the fish is broiled over a charcoal fire. After broiling, the fish is dried on a drying shelf.

Broiled-dried sardine is prepared as follows: small fish such as sardine are broiled near a charcoal fire and are then dried. After drying the bamboo stick is removed from the fish. Other fish are also prepared as broiled-dried foods by the same process.

Dried-Salted Products
Dried-salted fish is also an important product among the dried products in Japan. The important dried-salted fish are sardines, pompanos (jack mackerel), saury, mackerel, cod, Alaska pollack and others.

Dried-Salted Sardines
In every place where sardines are caught in Japan, dried-salted sardines are also processed. 'Ma-iwashi' (*Sardinopus melanosticta*) and round herring ('urume-iwashi', *Etrumeus micropus*) are used for the preparation of dried-salted products. There are three kinds of dried-salted sardines according to the method of preparation: round-drying, split-drying and pierced-drying.

a. Round-drying. The round herring ('urume-iwashi') makes a product superior in quality to that obtained from the sardine ('ma-iwashi'). Raw sardines are placed in a basket and washed in a sea water tank. Washed fish (15–16 kg) plus about 15% of salt are put into a barrel and the fish are covered with water. The fish in the barrel are stirred two or three times. An alternative method consists of putting the raw sardines into a salt solution of Bé 15°. The salting continues for 6–8 h. After salting, 10 fish are pierced from mouth to gill slit with a bamboo stick and dried in the air. The yield of the round dried-salted sardine is 20–30% of the raw sardine by weight.

b. Split-drying. Split-drying is employed to prepare the dried-salted product from the round herring ('urume-iwashi'). If sardines ('ma-iwashi') are employed as the raw material those of low fat content are preferable. The fish bodies are split at the ventral or dorsal side, and the gills and visceral mass removed. After having been washed, the fish are soaked in a salt solution of Bé 15° to 18°, or they are dry-salted overnight with 20–30% of salt. The salted fish are washed with fresh water and placed skin-side down on a bamboo blind and dried in the air. When the fish are two-thirds dried, they are turned over to continue until they are completely dried.

c. Pierced-drying. A bamboo stick is pierced through the eyes of sardine or from gill slit to mouth. As raw material, sardines or anchovies are used. After the fish are washed with fresh water, they are soaked in salt solution of Bé 6° to 7° for 4–5 h. The salted fish are pierced with a bamboo stick through the eyes, or from gill slit to mouth. The pierced fish are dipped in fresh water to wash, and are hung from wooden bars to dry for 5–6 days. The dried fish are packaged, five sticks in each bundle.

Dried-Salted Pompanos (Jack Mackerel) ('Shio-Boshi Aji')

Pompanos, which are caught in Japan in great abundance, are also salted and dried. There are two kinds of dried-salted pompanos; dried-split and pierced-dried. In Japan, splitting is carried out more often from the ventral side of the fish than from the dorsal side. The visceral mass is removed from the split fish, and they are washed with fresh water. After washing, the fish are dry-salted overnight and then are dried on a bamboo blind. The yield is 50–70% of the raw fish by weight.

The 'kusaya-no-himono' is a rather special product and production is limited. Scad (*Decapterus muroadsi*) are used as the raw material. The brine in which the scad are salted is used over and over again, and contributes a special putrefactive odour. Whenever the fish are put in the brine, salt is added little by little. The fish are soaked in the brine overnight. The next day, after draining the brine, the fish are dried in the sun. The special odour is liked by some people.

Dried-Salted Saury and Mackerel

Dried-salted saury and mackerel are prepared as described above for sardines. The main places of production of the former are Ibaragi, Chiba, Tokyo, Shizuoka, Mie, Kyoto and Hyogo Prefectures, and Chiba and Shizuoka Prefectures in the case of mackerel.

Dried Cod

The processing of dried cod in Japan is as follows: raw cod is split at the dorsal side and the visceral mass is removed. The fish is washed with fresh water to remove the black membrane lining the body cavity. In old Japanese processing the head was not removed. The split cod is salted in brine or with dry salt. When brining is employed, the split cod is soaked in salt solution of Bé 18° which had been boiled and then cooled. At this time, cod are piled on each other in an empty tank and the brine is added. After 2 days' soaking, the brine is renewed and the fish are pressed with a plate, on which stones are placed, for 1 day. The next day the brined fish are placed to dry on a bamboo blind. When dry-salting is employed, split cod are piled on each other with a sprinkling of fine salt between each two layers. The amount of salt used is about 190–200 kg for 1000 split fish. The salting continues for 10 days. After salting the fish are washed with salt solution of Bé 4° and placed on a bamboo blind to drain. After draining, the fish are dry-salted again in a tank. After washing the surface of the salted fish with fresh water the fish are again dried on a bamboo blind.

Dried Cod Fillet ('*Suki-Mi*')

Cod or Alaska pollack fillets are dry-slated or salted in brine and dried. After the fish are filleted, the fillets are skinned. Before filleting, head, bone and fins are removed. The salting is done in the same way as for dried round or split cod.

Dried-Salted Yellowtail ('*Enkan-Buri*')

Yellowtail ('buri', *Seriola quinqueradiata*) is used not only as a daily dish, but also as canned fish or dried-salted fish. The kinds of dried-salted yellowtail include big yellowtail ('enkan-buri'), small yellowtail ('enkan-inada'), small yellowtail fillet and yellowtail bound with straw. The amount of production is small, because most of them are used fresh.

a. Dried-salted yellowtail ('*enkan-buri*'). The visceral mass and gills are removed from fresh big yellowtail by splitting the ventral side. After being washed with fresh water, several lines of half-cuts are made on the surface of the fish body in order to allow salt to penetrate easily into the meat. The dressed fish is then soaked in dilute salt solution to remove blood and other extraneous materials from the meat. For each barrel, 10 big yellowtail are treated with a sprinkling of salt, to the amount of 15 to 16 kg. After salting for 4–5 days, the fish are washed with fresh water and are dried in the sun. Sometimes the big yellowtail are split from the dorsal side, and then are salted and dried.

b. Dried-salted yellowtail bound with straw (‘*waramaki-buri*’). After the head is cut off, the visceral mass is removed from the split ventral side of fresh big yellowtail. The dressed fish are washed with fresh water and filleted. About 22 kg of salt are added to each 10 fish (20 fillets). After dry-salting for about 1 week, the fish are washed with fresh water. Each fillet is cut lengthwise into two pieces. The pieces are hung for 1–2 weeks in the shade to dry. Fat which appears on the surface of the meat is removed with a bamboo spatula. The surface of the fish meat is covered with a thin layer of baked salt or rice-bran (1 kg) to which salt (500 g) is added. Each fish meat piece is wrapped with Japanese paper, then with straw and finally is bound tightly with straw rope. At the top of the rope, a small loop is made for hanging. Sometimes the bound fish is smoked to prevent oxidation.

c. Dried-salted small yellowtail. The head of the fish is cut diagonally above the eyes to the throat and the fish is then filleted after removal of visceral mass and gills. Each fillet is partially skinned. On the skinned meat three or six half-cut lines are drawn. After the skinned fillet is washed in dilute salt solution, about 500 g of salt is added to each two fillets (one fish) in a barrel. After 1–1½ days of salting, the fish fillet is dried in the sun for 7–10 days. The dried fillets are then packed 50 to a box.

Frozen-Dried Products

Alaska pollack are frozen outdoors during winter nights; the water in fish meat evaporates after being thawed in the sun in the day time, and drying is completed by repeated freezing and defrosting. These frozen-dried fish are called ‘mintai’ in Korea.

Frozen-Dried Alaska Pollack (‘*Mintai*’)

The ventral side of Alaska pollack is split to remove the visceral mass. Each 10 fish are bound together through the gill slit and they are soaked in fresh water to remove blood and to let the meat absorb sufficient water, the water being changed four to five times a day for 2 days. After this soaking, the bound fish are hung from wooden bars outdoors to freeze completely. The frozen fish are always placed outdoors in order to be dried by repeated freezing and thawing. When the drying is 80% complete, the bound fish are piled up on each other to the height of 1 m and are covered with a straw mat for 3–4 days to equalise their moisture content. In good weather, the bound fish are hung from wooden bars and dried outdoors. It requires about 70 days to complete the drying.

Fish-'Fushi' (Dried Fish Sticks)

Dried Bonito Sticks (*'Katsuo-bushi'*)

Fish-'fushi' are a special product in Japan. They are prepared by boiling, broiling (smoking) and drying fish fillets. As raw materials for the fish-'fushi', bonito, albacore, big eye tuna, mackerel and sardine are used. The fish-'fushi' are extensively used in preparing soups and broths. The fish-'fushi' are shaved with a sharp knife or blade and the thin flakes are added to hot water to make a soup base.

The fresh bonito are beheaded, eviscerated and filleted. The two fillets of small bonito (less than 3 kg in weight) are called 'kame-fushi'. The two fillets of bonito larger than 3 kg are again cut lengthwise into four fillets; the dorsal side fillets are called 'o-fushi' (back meat 'fushi'), and the ventral side fillets are called 'me-fushi' (belly meat 'fushi'). The back meat 'fushi' which is called 'o-fushi' is the most expensive.

The cut bonito meat (fillets) are placed on a tray for boiling. They must be placed in a position so that they remain straight. Eight to ten trays are placed in a round kettle, of which the diameter is about 1 m. The temperature in the kettle is 80–85°C for fresh bonito meat, and 90–95°C for stale bonito meat. The temperature is raised slowly so as to prevent splitting, for about 20 min to boiling. The boiling is continued for 45–60 min for 'kame-fushi', and for 60–90 min for 'o-fushi' and 'me-fushi'. Oily strips are boiled for a longer time than the usual ones. After boiling, the trays are taken from the kettle and placed into a large tub filled with water, in order to cool. During the cooling, the surface of 'kame-fushi' or 'me-fushi' is two-thirds skinned and small bones are removed.

The skinned and boneless fillets, after boiling, are placed on wooden trays with open-work bottoms. Groups of six to seven wooden trays are dried at first in the open air. When partially dried, the trays are placed in the broiling furnace (smoking furnace). These wood burning furnaces are constructed of stone or concrete and have an open top above which six or seven trays containing fillets are piled so that the smoke passes up through the layers in the tray. The distance between the lowest tray and the fire is about 1·4 m. Drying is continued until the dried surface becomes yellowish brown in colour. The trays are rotated in order to achieve uniform drying.

After the first broiling (smoking), cracks on the surface of the dried fillet are pasted with mashed waste meat of bonito which was boiled and broiled separately with the bonito fillets. The fixed fillets are placed back on the trays and dried in the broiling furnace again. The temperature of the second broiling is high enough to cause the fillet surface to lose moisture. At night the trays are taken from the furnace and are piled so that the water content in the fillet becomes uniform. The broiling (smoking) is repeated about ten times until the fillets have hardened. With the advance of the drying, the temperature of the furnace is lowered and the time of the broiling is shortened.

The broiled fillets are then dried as they are in the tray or on straw mats in the sunshine. When the dried fillets are placed in a box for 3–4 days, the surface of the dried fillet becomes soft. The softened fillets are shaped by shaving the surface with knives having various shapes. The shaved dried fillets are dried in the sunshine for another 2 days turning the back and the ventral sides.

When the dried fillets are placed in a box for 13–14 days, moulds (mainly *Penicillium* and *Aspergillus*) grow on the surface. This is called the first mould. The dried fillets having moulds are taken from the box and are dried at first in the shade, and then in the sunshine. After the drying thick layers of moulds, or undesirable moulds, are removed from the surface with a brush and the fillets are then placed in another box to allow the moulds to grow. This is called the second mould. The growing of mould is repeated four times. With an increase in the number of moulds, the colour of the surface of the dried fillet changes from greyish-green to light brown which is the characteristic colour of the dried bonito sticks. The purposes of the mould growth is to make a smooth surface, to adjust the moisture content, to decrease the apparent fat content, and to increase the desired taste. Some manufacturers employ the kind and growth of moulds as an index of quality of the dried bonito sticks. The good species of moulds grown on the surface of the dried bonito sticks are *A. glaucus*, *P. glaucum* and *A. melleus*. The undesirable species are *A. flavoviridescens*, *Torula* sp., *Cladosporium herbarum* and *Catenularia fuliginea*.

The yield of the finished production proportional to the weight of raw bonito is small, about 20 %.

During the storage of the dried bonito sticks, insects are likely to penetrate into the dried meat, making holes. Manufacturers sometimes suffer considerable loss because of insect damage. In order to prevent such damage, the bonito-'fushi' are sometimes dried in the shade, or are stored in cold storage at 0°C. Sometimes they are fumigated with carbon disulphide. Insecticides cannot be used because they inhibit the surface mould growth.

Seasoned-Dried Fish

Seasoned-dried fish has been processed in Japan from ancient times, and is a regular item in the daily menu. This product is eaten after being broiled. Fish are dressed and seasoned with a seasoning solution made from sugar, 'shoyu' and sweet 'sake' and are then dried in the sun.

Small fish are used for the preparation of this product: e.g. sardine, small sea-bream, small flat fish or pompanos (jack mackerel), which have low fat content. Small fish are split at the ventral side for removal of the viscera. Medium sized fish are filleted. The dressed fish are soaked in the seasoning solution for 1–2 h. The seasoning solution is made from 'shoyu' (1·8 litre), sugar (45–560 g), sweet 'sake' (360 ml) and sodium glutamate (3·75 g). In order to give brilliance to the surface of the finished product a slight amount of agar, gelatine, wheat-gluten or glucose is added to the seasoning solution. The seasoned fish are dried in the sun.

EFFECT OF SMOKING AND DRYING ON CHEMICAL COMPOSITION OF FISH

Chemistry of Smoked and Dried Fish

References to much of the Japanese work on the effects of drying and smoking on proteins, lipids and vitamins have been incorporated into Chapters 2, 3 and 4 respectively. The following sections are intended to supplement the general findings and to emphasise the importance of the Japanese literature to this topic.

Yields and Water Contents

During smoke-drying fish gradually lose weight as water is lost. On the one hand, lowering water content gives a higher preservability to the product, but on the other it makes the meat tough and reduces the yield.

Yoshikawa and Tamamoto (1942) observed changes in the weight of herring during the processing of smoked fish. The result is shown in Table 1. In Table 1, the weight loss is approximately 10% after the curing and

TABLE 1

CHANGE IN THE WEIGHT OF HERRING DURING SMOKING

Herring	*Fresh fish*	*After curing* (1 *week*)	*After air-drying*	*After smoking* (1 *day*)	*After smoking* (3 *days*)	*After smoking* (10 *days*)	*After smoking* (15 *days*)
Large size							
Weight (g)	240	227·7	216·3	189·5	170·3	135·0	122·9
Per 100g fish	100	94·9	90·1	78·8	70·9	56·2	51·2
Medium size							
Weight (g)	215	201·2	159·9	163·9	141·5	105·3	91·7
Per 100g fish	100	93·4	90·9	76·0	65·6	48·8	42·5
Small size							
Weight (g)	199	188·6	183·4	150·3	128·0	90·9	77·5
Per 100g fish	100	94·5	92·0	75·4	64·1	45·5	38·8

TABLE 2

CHANGE IN PROXIMATE COMPOSITION OF HERRING DURING SMOKING

Composition	*Fresh fish*	*After smoking (5 days)*	*After smoking (10 days)*	*After smoking (15 days)*
Water (%)	77·50	58·25	44·50	36·50
Protein (%)	14·30	28·37	33·94	37·43
(dried matter)	63·55	67·95	61·15	58·94
Fat and oil (%)	5·35	6·95	8·45	14·50
(dried matter)	23·77	14·02	15·22	22·83
Ash (%)	2·21	10·35	13·68	15·34
(dried matter)	9·82	24·79	24·64	24·15
NaCl (%)	0·92	9·50	9·70	9·60
(dried matter)	4·09	22·77	15·67	15·11
Water activity	0·98	0·89	0·85	0·82

air-drying processes. The rates are approximately 5% per day for the first 5 days, then less than 5% per day during the subsequent weeks. The weight loss is larger in small fish than in larger ones.

Yoshikawa and Tamamoto (1942) also measured the chemical composition of herring during processing smoked products, as shown in Table 2. In the fish, the water content as well as the weight decreases during the smoking process: the proportions of other constituents increase conversely. The water content after smoking for 15 days is approximately half that of the fresh fish.

Digestibility

Although the work done by Kajimoto and Tanita (1962) was not on smoked and dried fish, it could have some relevance here. They pointed out that the digestibility of whale meat fried with oxidised oil, of which the peroxide value was greater than 200, was poorer than that of whale meat which had been fried with non-oxidised oil. Further, Ito (1970) compared the digestibilities of fresh and fried carp, and noted that fresh carp was most digestible, which was followed by lightly and heavily fried fish in that order.

Nitrogenous Compounds

Tadokoro and Watanabe (1928) made a study regarding the denaturation of muscle proteins in salmon by smoke-drying and found that: (a) the salt-

TABLE 3

CHANGE IN MUSCLE PROTEIN CONTENTS OF HERRING DURING SMOKE DRYING (g PER 100 g DRIED MATTER)

	Fresh fish	*After smoking (5 days)*	*After smoking (10 days)*	*After smoking (15 days)*
Crude protein	83·39	81·53	80·64	78·43
Water soluble protein	17·73	15·38	13·29	8·86
{ Protein	3·54	3·31	3·21	3·21
{ Amino-N	1·40	1·21	1·22	1·22
{ Coagulable proteins	0·088	0·088	0·088	0·088
Insoluble protein	65·66	66·15	67·35	69·57

TABLE 4

NITROGEN DISTRIBUTIONS IN FRESH AND SMOKED HERRING (g N per 100 g total N)

	Ammonia	*Humine*	*Mono-amino*	*Diamino*	*Arginine*	*Histi-dine*	*Lysine*	*Amide*
Fresh	1·02	0·79	58·04	39·50	15·70	6·49	16·77	1·67
Smoked	1·50	2·26	57·96	39·64	16·78	5·91	15·92	0·16

soluble proteins such as myosin decreased, while insoluble and non-protein nitrogenous compounds increased; (b) ash and phosphorus contents were low in myofibrils isolated from the muscle of smoked salmon; (c) ammonia and diamino nitrogenous compounds such as arginine and histidine decreased, while monoamino nitrogen increased and (d) free amino nitrogen increased. Changes were found in several types of nitrogen in the muscle proteins of herrings by smoke-drying (Yoshikawa and Tamamoto, 1942).

In Table 3, it can be seen that the water soluble proteins are reduced during the process of smoke-drying. Table 4 shows the distribution of nitrogen in fresh and smoked herring (Yoshikawa and Tamamoto, 1942). Decomposition of muscle proteins results in an increase of ammonia and humine and a decrease of amide by smoke-drying fish.

Fukuhara and Kuroda (1950) observed chemical changes in herring

TABLE 5

CHANGE IN NITROGENOUS COMPOUNDS OF HERRING DURING SMOKE DRYING (G N PER 100 G FRESH MEAT)

Materials	*Total*	*Protein*	*Coagulable*	*Peptone*	*Ammonia*	*Free amino*	*Organic matter*	*Inorganic matter*
Fresh muscle	0·53	0·33	0·23	0·18	0·02	0·04	3·43	1·15
Salted muscle	0·58	0·45	0·25	0·06	0·03	0·08	6·02	4·26
After smoking (7 days)	0·85	0·52	0·47	0·08	0·05	0·08	6·95	5·55
After smoking (14 days)	1·38	0·93	0·61	0·14	0·10	0·19	10·18	7·30
After smoking (21 days)	1·76	1·17	0·72	0·17	0·13	0·22	13·12	9·28
After smoking (28 days)	2·24	1·70	0·88	0·24	0·14	0·31	13·44	10·46
Commercial product	1·42	0·59	0·29	0·14	0·41	0·15	13·18	4·92

during cold smoking and obtained the results shown in Tables 5 and 6. Significant increases were observed in both nitrogenous and soluble compounds in herring with progressing smoke-drying. At a stage when smoke-drying had progressed to some extent, approximately 45% of the weight of the muscle was converted to soluble compounds, in which about 30% were organic substances and the remainder was sodium chloride. In the fish, more than 40% of the nitrogenous compounds were rendered soluble. From this observation it can be presumed that autolytic or other reactions taking place in the muscle can break down nitrogenous compounds to produce free amino acids during smoke-drying and this leads to an improvement in the taste of smoked herring.

Migita *et al.* (1956) reported that salt soluble proteins, such as myosin, were denatured by dehydrating fish, while water soluble proteins, such as myogen, were not.

Throughout the processing of smoked fish the nitrogenous content of the fish varies with different steps. The nitrogenous contents increase in herring muscle, when the fish contain salt at a concentration of less than 1% during the smoking process (Akiba *et al.*, 1966). Akiba *et al.* (1967) also reported that the amino and volatile basic nitrogen contents of

TABLE 6

CHANGE IN THE CONTENTS OF WATER AND NITROGENOUS COMPOUNDS OF HERRING DURING COLD SMOKING

Materials	*Water content (%)*	*Total N (%)*		*VB N (%)*[a]		*Amino N*	
		In wet matter	*In dry matter*	*In wet matter*	*In total N*	*In wet matter*	*In total N*
Fresh muscle	77·30	1·84	8·11	0·0163	0·88	0·083	4·51
Salted muscle	77·25	2·26	8·14	0·0174	0·77	0·097	4·28
After air drying	70·10	2·45	8·20	0·0182	0·75	0·121	4·49
After smoking	65·20	2·84	8·17	0·0262	0·92	0·174	6·13
After smoking (3 days)	61·30	3·21	8·28	0·0345	1·08	0·203	6·33
After smoking (5 days)	58·25	3·44	8·24	0·0433	1·19	0·215	6·25
After smoking (7 days)	54·40	3·78	8·30	0·0453	1·15	0·220	5·82
After smoking (10 days)	50·30	4·10	8·24	0·0483	1·11	0·223	5·49
After smoking (12 days)	43·20	4·67	8·23	0·0539	1·03	0·235	5·03
After smoking (13 days)	40·25	4·96	8·28	0·0580	1·09	0·243	4·91
After smoking (15 days)	39·20	5·02	8·39	0·0664	1·32	0·324	6·45

[a] VB N = volatile basic nitrogen.

herring increase during the smoking process if the relative humidity is between 80 and 90%.

Konosu and Hashimoto (1959) studied changes in the free amino acids during the manufacture of 'katsuobushi' (dried bonito) and pointed out features as follows: (a) histidine was about 80% of total free amino acids; (b) the contents of taurine, lysine and alanine were also high; (c) a portion of each amino acid was diminished at the boiling process, and (d) during smoking, some amino acids increased in concentration but others stayed at almost the same levels. As to the solubility of phosphorus and nitrogen, Nakane (1959) reported that 13% of the soluble nitrogenous compounds is dissolved in hot water by boiling for 15 min and 16% for 5 min, while 69% of soluble phosphorus compounds is dissolved by boiling for 15 min and 78% for 5 min. These findings suggest a considerable variance of nitrogen content in the meat before and after smoking. It may be concluded, therefore, that curing as a process leads to a reduction in the nitrogen content of fish while smoking leads to an increase

Some papers dealt with the amounts of nitrogenous compounds and the chemical composition in fish and the various factors involving processing. Isa (1961*a*, *b*) has reported that while no such tendency was observed in nitrogenous material, including amino acids, they vary in amount according to salt content in boiled and dried anchovy; the lower the total nitrogen is, the higher the salt content of the product. Also, the amount of extractive nitrogenous material, including amino nitrogen, varies somewhat in proportion to the oil content of boiled and dried anchovy. Positive correlation could not be found between the oil and salt contents.

Igawa *et al.* (1965) conducted a study on changes in the chemical composition of scallops by heating and learned that the extents of dehydration rates were 26·75 % at 80°C for 25 min and 44·23 % at 100°C for 15 min. The total nitrogen content was found to decrease gradually by heating at 60°C and 80°C, but to decrease rapidly at 100°C. The higher the heating temperature, the faster the glycogen dissolved into the brine. Konosu and Matsuura (1960) reported, however, that the amount of tryptophan in raw fish flesh ranged from 1·25 to 1·35 % and that variations in the contents between the species of fish was quite small. They also mentioned that the tryptophan content of canned and dried-salted fish muscle was almost at the same level as that of raw flesh.

There are some papers which reported on the amino acid composition of cooked or dried fish. Amino acids detected in extractives from heated sea-bream were aspartic acid, glycine, alanine, leucine, glutamic acid, valine, histidine, threonine, tyrosine, asparagine and proline (Yabuki, 1965). Ando (1965) has studied amino acids in dried-salted sand-fish, and reported that glutamic acid, alanine, leucine and lysine contents were high and sulphur amino acids and tryptophan contents were low.

When the raw material deteriorates, however, the content of nitrogenous compounds becomes variable and such organic bases as trimethylamine, histamine, arcaine, putrescine, agmatine, cadaverine and iso-amylamine can be detected in dried squid (Tanikawa and Motohiro, 1958). In this case, the compounds may be unfavourable not only for human health, but they may also have a bad effect on the food quality of dried products. A white powder produced on the surface of dried squid, giving it a poor appearance, is known to consist of some nitrogenous compounds such as betaine, taurine, aspartic acid, glutamic acid, histidine and threonine which are derived from the decomposition of squid proteins (Tanikawa *et al.*, 1953).

Some papers are concerned with non-enzymatic browning discoloration. The reaction is a rather important aspect of food quality, but also relates to nutritive value of the product. The development of the brown discoloration was negligible by reaction of the non-volatile fraction of amino compounds from boiled-dried anchovy, dried mackerel and salted trout with autoxidised methyl linoleate and the methyl ester of cuttlefish liver oil; it was markedly positive, however, by a reaction with the volatile fraction (Fujimoto *et al.*, 1968).

The browning of fish flesh by heating has been known to involve sugar–amino (protein) reactions; the loss of sugars in the flesh by autoclaving was more or less greater in the chopped state than in the block, but there was no clear relationship between the state of flesh and the degree of browning (Nagayama, 1960). The level of total acid-soluble hexose was decreased by heating in browned flesh. Inorganic phosphorus in the flesh was increased by heating. Total and soluble arginine, and total lysine in the flesh were decreased, whereas soluble lysine was increased (Nagayama *et al.*, 1962).

On the browning of dried fish products, crotonaldehyde and acetaldehyde both enhance browning, while ribose shows no influence (Koizumi *et al.*, 1959). Kajimoto (1961) studied the browning discoloration of smoked cod and octopus and reported that hydroquinone contained in smoke is converted to quinone by reacting with oxygen in association with ammonia formed from raw fish. The quinone is then changed to hydroxyquinone which can polymerise to form a black-coloured high molecular weight compound.

Ribonucleic Acid and its Derivatives

The work done on ribonucleic acid and its derivatives reflects the importance of flavour and discoloration to food quality. Ueoka *et al.* (1970) reported that s-RNA, DNA and r-RNA in fish meat were not changed by the freezing, drying and rehydration processes.

Boiled, dried and canned fish generally showed a high inosinic acid content due to inactivation of the enzyme system connecting with degradation of inosinic acid by heat. The change of inosinic acid content during drying and autolysis was followed using jack mackerel meat. A considerable amount of the acid was retained in dried fish (Fujita and Hashimoto, 1960).

By evaluating the effect of drying methods of mackerel on inosinic acid content, Shimoda *et al.* (1975) reported that inosinic acid content

TABLE 7

CHANGE IN MICROFLORA DURING COLD SMOKING OF HERRING

Days for smoking	*Detection of bacteria*			*Number of species*	*Water content* (%)
	Gill	*Intestine*	*Back bone*		
1 day	++	++	++	6	62·8
1 week	–	+	+	3	52·6
2 weeks	–	–	+	2	50·0
3 weeks	–	–	+	3	38·2

decreased to 50% during processing. Continuous heating resulted in a low concentration of the acid.

In connection with the orange discoloration of canned skipjack, changes in the amounts of glycolytic intermediates in skipjack flesh during thawing and cooking have been studied (Yamanaka *et al.*, 1974). They found that *d*-glucose 6-phosphate accumulated during thawing and the early stages of cooking but it had disappeared completely after the double cooking (heating for 100 min at 100°C followed by 90 min at 113°C).

In preparations for smoked squid, the flavour of the product could be improved by enrichment with 5′-ribonucleotide and disodium succinate (Tanikawa *et al.*, 1963).

Microbiology of Smoked and Dried Fish

Changes in Microflora during Smoking

During the initial stage of smoking, chemical compounds contained in the smoke hardly penetrate into fish flesh; they are distributed only on the surface. After smoking the fish for 1 week, few bacteria could be observed at the gill but penetration of the chemical compounds into the belly part was slight. After smoking the fish for 2 weeks, the chemical compounds were detectable almost throughout the body except in the centre of the fish; in this case bacteria were rarely observed. Changes in microflora during fish smoking are shown in Table 7 (Sasa, unpublished data).

It was reported that exposing bacteria to smoke for 2·5 h effectively killed *Bacillus anthracis*, *Staphylococcus aureus* and *Streptococcus pyogenes*; exposure for 3 h killed *B. subtilis* and for 18 h the spores of *B. anthracis*.

TABLE 8

COMBINED EFFECT OF SALMINE SULPHATE AND POTASSIUM SORBATE ON THE GROWTH OF MOULDS AT pH 6·0[a]

Strain	*Concentrations of salmine sulphate and sorbate (%)*				
	0·06(p) + 0·04(s)	0·07(p) + 0·03(s)	0·08(p) + 0·02(s)	0·09(p) + 0·01(s)	0·1(p)
Eurotium echinulatum	+	+	+	+	+
Eurotium repens	+	+	+	+	+
Eurotium amstelodami	+	+	+	+	+
Eurotium rubrum	+	+	+	+	±
Botryotinia fukelina	+	+	+	+	−
Epicoccum purpurascens	+	+	+	+	+
Penicillium digitatum	+	+	+	+	+
Aspergillus conicus	+	+	+	+	+
Rhyzopus oryzae	−	−	+	+	+
Rhyzopus stolonifer	−	−	+	+	+
Wallemia sebi	−	−	−	−	−
Myceliophthora lutea	+	+	+	+	+
Mucor circinelloides f″ circinelloides	−	+	+	+	+
Mucor racemosus f″ racemosus	−	+	+	+	+
Aspergillus niger	−	−	−	−	−

[a] Agar plate method.
+ = growth.
± = growth inhibited.
− = resistant.
(p) = protamine (salmine) sulphate.
(s) = potassium sorbate.

Microflora in Dried and Smoked Fish Products

In general, the preservability of dried and smoked fish is induced by the dehydration process. The water activity (Aw) of smoked fish is approximately 0·9, and at this level some microbial genera such as *Micrococcus, Staphylococcus, Sarcina* and fungi will grow. Products deteriorate by the growth of moulds if the water content is approximately 15% (Kaneko, 1976). *Penicillium, Aspergillus, Mucor* and *Fungi imperfecti* predominated in isolates from smoked squid (Sato and Okuda, 1963), scallops and cod (Sato and Okuda, 1964).

In Japan, sorbic acid or potassium sorbate may be used as food additives for smoked squid and octopus and for dried fish and shellfish up to the levels of 1.5 g/kg and 1 g/kg, respectively. Tanikawa *et al.* (1964) found that the proportion of sorbic acid present in the final product was 65–88 % of the added amount through processing squid. At the final stage of processing, the sorbic acid could be reduced, by heating, to levels varying from 10 % to 60 % of the initial amount. The finding suggests that an effective use of sorbic acid is after the final stage of the processing for smoked squid.

From the point of view of food safety, however, it is a recent trend that the levels of such synthetic preservatives as sorbic acid or sorbates have to be reduced in the food to levels as low as possible, even when their addition to specified kinds of foods is permitted below certain limits of concentration. Protamines, including salmine and clupeine, are effective bactericides and fungicides, and are put in the category of natural preservatives (Islam *et al.*, 1984; Kamal *et al.*, 1986). Kamal and Motohiro (1987) studied the synergistic effect of increasing levels of potassium sorbate with varying concentrations of salmine sulphate on the growth of spoilers from dried fish; the results are shown in Table 8.

Combinations of 0·08 % salmine sulphate plus 0·02 % sorbate and 0·09 % salmine sulphate plus 0·01 % sorbate were found to be most suitable for inhibiting the growth of most mould species. The pH greatly influenced the survivability of spores in the presence of both the preservatives. The combination of 0·08 % salmine sulphate plus 0·02 % sorbate, however, exerted lethal effects on conidial spores over a wide pH range.

Kamal and Motohiro (unpublished data) have also studied the synergistic effect of salmine sulphate with ethanol on the growth of moulds isolated from spoiled dried fish. Of the various combinations, 0·05 % salmine sulphate plus 5·0 % ethanol was the most efficient, inhibiting most of the organisms. All the organisms were inhibited by 0·1 % salmine sulphate plus 6·0 % ethanol, whereas 8·0 % ethanol was necessary for total growth inhibition.

In dried fish, spoilage caused by growth of moulds is noticeable. The species involved are *Aspergillus giganteus, A. gracils, A. restrictus, A. tubingensis, A. versicolor, Chaetomium cochlicrides, Epicoccum purpurascens, Eurotium amstelodami, E. repens, E. herburiorum, Penicillium expansum, P. olivino-viride, P. viridicatum* and *Phoma* sp.

The growth of moulds during the production of dried bonito sticks affects the quality of the final product but pure cultures of the moulds are seldom introduced. The number of species which can grow on the surface of half-dried material is limited. At the final stage of the production, the species of so-called 'moulds on dried bonito stick' include *Eurotium repens* and *E. rubrum*, and these produce a particular flavour. The species found in the final product occasionally include *Aspergillus ochraceus*, *A. oryzae*, *A. ostianus*, *A. tamarii*. *Penicillium cyclopium echinulatum*, *P. putterilli* and *Syncephalastrum racemosum* (Udagawa and Tsuruda, 1975).

Obata and Matano (1953) pointed out that a particular flavour produced by growing *Eurotium repens* and *E. rubrum* upon fish surface relates closely to fish oil. They successfully synthesised a dried bonito stick flavour by incubating *E. repens* and *E. rubrum* with casein or defatted soya beans onto which mackerel oil was brushed.

The preservability of dried-salted squid depends on dehydration by salting, but not on salt content (Motohiro and Fukushima, 1959). Since the water content affects the preservability of dried fish, rehydration of the product by some reason may cause microbial deterioration.

Tanikawa *et al.* (1965) reported that the sugar and extractives exuded from cod by leaving the product in a very humid environment may be a medium for growth of mould.

REFERENCES

Akiba, M., Motohiro, T. and Kudo, S. (1966). Studies on smoked fish product III. Manufacture of smoked herring (2). *Bull. Fac. Fisheries, Hokkaido Univ.* **17**, 157–61.

Akiba, M., Motohiro, T. and Kudo, S. (1967). Studies on smoked fish product IV. Manufacture of smoked herring (3). *Bull. Fac. Fisheries, Hokkaido Univ.* **17**, 229–33.

Ando, K. (1965). Amino acids in food IV. Amino acids in salted dried sand fish. *Report Hokkaido Inst. Public Health* **12**, 151.

Fujimoto, K., Murayama, M. and Kaneda, T. (1968). Studies on the brown discoloration of fish products I. Factors affecting the discoloration. *Bull. Japan. Soc. Sci. Fisheries* **34**, 519–23.

Fujita, T. and Hashimoto, Y. (1960). Inosinic acid content of foodstuffs IV. Marine products. *Bull. Japan. Soc. Sci. Fisheries* **26**, 907–10.

Fukuhara, T. and Kuroda, K. (1950). Change in herring meat during smoke drying. *Monthly Report, Hokkaido Fisheries Exptl Sta.* **7**, 17–23.

Igawa, K., Kuroda, K. and Ichisugi, T. (1965). Changes in chemical compositions of scallops by heating. *Monthly Report, Hokkaido Fisheries Exptl Sta.* **22**, 327–36.

Isa, Y. (1961*a*). Studies on the boiled and dried anchovy II. Possible correlation among some chemical components of boiled and dried anchovy. *Bull. Japan. Soc. Sci. Fisheries* **27**, 1075–9.

Isa, Y. (1961*b*). Studies on the boiled and dried anchovy III. On crude fat content among three body portions of head, viscera, and muscle and presence of the so-called oily anchovy. *Bull. Japan. Soc. Sci. Fisheries* **27**, 1080–3.

Islam, N.M., Itakura, T. and Motohiro, T. (1984). Antibacterial spectra and minimum inhibition concentration of clupeine and salmine. *Bull. Japan. Soc. Sci. Fisheries* **50**, 1705–8.

Ito, K. (1970). Changes in digestibility of fish meat proteins by different cooking methods IV. Fried fish. *Report Tottori Univ. Educ. (Nutr. Sci.)* **21**, 96–102.

Kajimoto, G. (1961). Effect of rancid oil on cooking VIII. Browning discoloration of smoked cod and octopus. *J. Japan. Soc. Food Nutr.* **14**, 122–9.

Kajimoto, G. and Tanita, H. (1962). Effect of rancid oil on cooking XIII. Protein denaturation and digestibility of whale meat fried with rancid oil. *J. Japan Soc. Food Nutr.* **14**, 467–71.

Kamal, M. and Motohiro, T. (1987). Combined effect of salmine sulphate and sorbate on the growth of moulds. *Nippon Suisan Gakkaishi* **53**, 867–72.

Kamal, M., Motohiro, T. and Itakura, T. (1986). Inhibitory effect of salmine sulphate on the growth of moulds. *Bull. Japan. Soc. Sci. Fisheries* **52**, 1061–4.

Kaneko, S. (1976). Smoked meat and microorganisms. *New Food Ind.* **18**, 17–23.

Koizumi, C., Kurobe, S. and Nonaka, J. (1959). On the browning of dried fish products. *Bull. Japan. Soc. Sci. Fisheries* **25**, 368–72.

Konosu, S. and Hashimoto, Y. (1959). Change of free amino acids during the manufacturing process of 'katsuobushi' (dried bonito). *Bull. Japan. Soc. Sci. Fisheries* **25**, 307–11.

Konosu, S. and Matsuura, F. (1960). Tryptophan content of fish meat. *Bull. Japan. Soc. Sci. Fisheries* **26**, 1040–9.

Koyama, H. (1960). Riboflavin contents in processed protein foods. *J. Japan. Soc. Food Nutr.* **12**, 325–31.

Migita, V., Matsumoto, J. and Saishu, T. (1956). On the denaturation of fish muscle proteins by dehydration. *Bull. Japan. Soc. Sci. Fisheries* **22**, 433–9.

Motohiro, T. and Fukushima, S. (1959). Studies on the complete utilization of squid XVIII. On the manufacture of salted dried squid. *Bull. Fac. Fisheries, Hokkaido Univ.* **9**, 311–28.

Murayama, S. and Yanase, M. (1960). Nutrients in fish meal produced in Japan and in other countries. *Report Tokai Regional Fisheries Res. Inst.* **27**, 55–61.

Nagayama, F. (1960). Studies on the browning of fish flesh II. Change of sugar content by heat process and browning. *Bull. Japan. Soc. Sci. Fisheries* **26**, 1026–31.

Nagayama, F., Hiraide, H. and Sano, K. (1962). Studies on the browning of fish flesh XI. Effect of heat on the levels of sugar, phosphate, arginine and lysine in flesh. *Bull. Japan. Soc. Sci. Fisheries* **28**, 1188–91.

Nakane, K. (1959). Dissolving of phosphorus and nitrogen compounds in dried-smoked bonito ('katsuobushi') into hot water. *Doshisha Women's Coll.* **10**, 355–61.

Obata, Y. and Matano, K. (1953). Studies on the flavors of katsuobushi I. On the factors affecting the development of the flavors. *Bull. Japan. Soc. Sci. Fisheries* **19**, 151–4.

Sato, T. and Okuda, Y. (1963). Moulds isolated from smoked squid. *Monthly Report, Hokkaido Fisheries Exptl Sta.* **1**, 61–5.

Sato, T. and Okuda, Y. (1964). Moulds isolated from dried scallops and seasoned-dried cod. *Monthly Report, Hokkaido Fisheries Exptl Sta.* **2**, 79–84.

Shimoda, Y., Waka, J., Hasegawa, K., Sawada, T. and Kawamura, M. (1975). Improving of food quality of dried-smoked fish III. Effect of drying methods of mackerel on inosinic acid content. *Report Shizuoka Fisheries Exptl Sta.* **1974**, 66–74.

Shinno, H., Kisaka, M. and Nakamura, S. (1965). Amino acids in muscle proteins of flat fish and digestibility by pepsin. *Report Food Kyoto Women's Coll.* **17**, 12–18.

Tadokoro, T. and Watanabe, S. (1928). Studies on denatured proteins-III. Denaturation of fish proteins. *J. Japan. Agric. Chem.* **4**, 1010–16.

Tanikawa, E. (1985). *Marine Products in Japan* (revised edn), Koseisha-Koseikaku Company, Tokyo.

Tanikawa, E. and Motohiro, T. (1958). Studies on the complete utilization of squid XVI. Amines produced in dried squid. *Bull. Fac. Fisheries, Hokkaido Univ.* **9**, 73–80.

Tanikawa, E., Kudo, D. and Motohiro, T. (1953). Studies on the complete utilization of squid IX. On the manufacture of dried squid (2). Chemical compositions of white powder on the product. *Bull. Fac. Fisheries, Hokkaido Univ.* **4**, 234–8.

Tanikawa, E., Akiba, M. and Motohiro, T. (1963). Studies on the complete utilization of squid XIX. Manufacture of smoked squid meat (1). *Bull. Fac. Fisheries, Hokkaido Univ.* **14**, 170–81.

Tanikawa, E., Akiba, M. and Motohiro, T. (1964). Studies on the complete utilization of squid XX. Manufacture of smoked squid meat (2). *Bull. Fac. Fisheries, Hokkaido Univ.* **14**, 243–61.

Tanikawa, E., Motohiro, T. and Akiba, M. (1965). The formation of white powder on seasoned dried cod. *Bull. Fac. Fisheries, Hokkaido Univ.* **16**, 189–200.

Udagawa, S. and Tsuruda, O. (1975). In: *Food and Moulds, an Introduction to Food Mycology*, Ishiyaku Publishers Inc., Tokyo, p. 236.

Ueoka, Y., Suemitsu, H. and Oka, H. (1970). Freeze dried fish VIII. Changes in nucleic acids in fish meat during freezing and drying. *Report Gen. Chem. Technol. Inst., Ehime Prefecture* **8**, 18–24.

Yabuki, Y. (1965). Amino acid contents in extractives from heated food II. Fish. *Mem. Nara Univ. Education* **13**, 67–73.

Yamanaka, H., Bito, M. and Yokoseki, M. (1974). Orange discolored meat of canned shipjack III. Changes in amounts of glycolytic intermediates in skipjack meat during thawing and cooking. *Bull. Japan. Soc. Sci. Fisheries* **40**, 941–7.

Yoshikawa, Y. and Tamamoto, I. (1942). Change in chemical compositions of fish during smoking process – I. Cold smoked herring. *J. Fisheries* **50**, 6–11.

Chapter 9

DRIED AND SMOKED FISHERY PRODUCTS: PREPARATION AND COMPOSITION

The following tables, with their associated notes, have been prepared by various members of the working party with the assistance of colleagues and others. They should help to give indications of the time and temperature regimes to which fish are exposed during the preparation of different smoked and dried products. In the instances where analytical data are absent it should be possible to draw conclusions, albeit limited ones, concerning the effects of processing on various aspects of nutritional value.

Additionally, where it has proved possible, the water activity values (Aw) for some of the products have been calculated using the compositional data provided in the formula given in Chapter 1 by Olley, Doe and Heruwati. These values will give an indication of the storage stability of the products in terms of fungal and microbial spoilage when they are compared with the three dimensional envelope illustrated in Fig. 6, chapter 1.

Some of the figures given in the tables have been culled from the literature and others, no less valid, have been received through the 'personal communication' route. In the interests of clarity, however, it was decided, with one exception,† that the sources of individual values would not be cited.

Products from the following countries or regions are described:

India (Mangalore) (Table 1)	Poland (Table 7)
Indonesia (Table 2)	Sri Lanka (Table 8)
Japan (Table 3)	Thailand (Table 9)
Malaysia (Table 4)	United Kingdom (Table 10)
Norway (Table 5)	West Africa (Table 11)
Philippines (Table 6)	Zambia (Table 12)

† Most of the figures in Table 9 are from *The Nutritive Value of Thai Foods* (1978), Department of Nutrition, Division of Public Health, Thailand.

TABLE 1

INDIA (MANGALORE): SMOKED AND DRIED FISHERY PRODUCTS

Product	*Species*	*Salting regime*	*Smoking and/or heating regimes*	*Composition (%)*			
				Salt[a]	*Fat*[a]	*Water*	*Protein*[a]
Salted and dried							
Anchovy	*Stolephorus* sp.	24 h		19·1	4·0–	24·0	58·0–
		20 %			8·0		65·0
Caranx	*Caranx* sp.	24–36 h		12·7–	5·1–	14·7–	55·1–
		25 %		28·4	8·4	25·7	69·3
Catfish	*Tachysurus* sp.	24–36 h		12·1–	3·4–	16·7–	58·9–
		25 %		18·7	9·2	43·6	74·2
Croaker	*Johnius* sp.	24–36 h		24·6–	6·9–	24·3–	45·9–
		25 %		34·3	13·7	45·7	61·3
Lactarius	*Lactarius*	24 h		23·4–	6·7–	18·7–	44·9–
	lactarius	20 %		34·7	11·7	31·4	66·1
Mackerel	*Rastrelliger*	48 h		11·3–	6·4–	21·0–	48·0–
	kanagurta	25–35 %		17·2	10·0	23·4	60·0
Rays	*Aetobatus* spp.	–		–	–	–	–
Ribbon fish	*Trichiurus* sp.	–		–	–	–	–
Sardine	*Sardinella*	24–36 h		16·8–	4·8–	28·4–	49·4–
	gibbosa	25 %		30·6	11·3	41·2	70·3
	Sardinella	24–36 h		16·9–	4·9–	16·9–	52·0–
	fimbriata	25 %		25·8	11·7	27·8	69·3
Shark	*Scoliodon* spp.	24–48 h		13·6–	2·5–	36·1–	44·0–
	Sphyrna sp.	25 % salt		25·0	3·6	44·93	52·0
Silverbelly	*Leiognathus* spp.	24 h		19·2–	2·4–	16·6–	48·3–
		20 %		35·2	19·1	43·6	67·1
Soles	*Cynoglossus* sp.	24 h		10·0–	1·7–	14·3–	60·3–
		20 %		26·1	4·9	35·2	75·4
White sardine	*Kowala coval*	24 h		16·1–	4·3–	16·5–	64·3–
		20 %		24·1	7·4	31·7	69·3

Salted and sun-dried							
Lizard fish	*Saurida tumbil*	–		–	–	–	–
Pink perch	*Nemipterus japonicus*	–		–	–	–	–
Seer Steaks	*Scomberomorus* sp.	–		–	–	–	–
Salted (undried)							
Ribbon fish	*Trichiurus* sp.	–		–	–	–	–
Dried							
Shrimp	*Acetis* sp.	–		–	–	–	–
Hot smoked							
Lizard fish Fish shavings	*Saurida tumbil*	2 h in 1:10 ratio of salt to fish (dry salting)	Smoking at 75°C till the moisture level reduced to 10%	5·6	10·50	4·50	73·00
Mackerel (Butterfly style)	*Rastrelliger kanagurta*	2 h in 1:8 ratio of salt to fish (dry salting)	Smoking for 5 h at 70°C ± 5°C in a modified traditional smoke chamber	4·90	14·00	45·00	36·00
Oil sardine Flakes	*Sardinella longiceps*	5 h in 1:6 ratio of salt to fish (dry salting)	Smoking at 45°C ± 5°C for 3 h and 75°C ± 5°C for 2 h	15·00	24·00	38·00	29·00
Pink perch Fish flakes	*Nemipterus japonicus*	Brining for 2–3 min	Hot smoking for about 4 h at 70°C and sun-drying to a level of 32% moisture	10·50	8·50	31·00	47·00
Tuna Fish shavings	*Euthynnus affinis*	60 min in saturated brine	Smoking at 70–80°C for 8 h and drying at 70°C for 16 h in hot air oven. Further dried till the moisture content is reduced to about 15% in sun or dryer	4·5	3·00	15·00	74·00
Tuna (Masmeen)	*Euthynnus* spp.	Fillets brining for 30 min	Hot smoking for 4–5 h	2·0	–	6·00	65·00

[a] Values are expressed on moisture-free basis.

TABLE 2

INDONESIA: SMOKED AND DRIED FISHERY PRODUCTS

Product	*Species*	*Salting, smoking and/or heating regimes*	*Composition (%)*					
			Salt	*Fat*	*Water*	*Protein*	*Ash*	*Aw*
Dried salted								
Lizard fish	*Saurida* sp.	25% salt for 37 h salting, 11·5 h drying at 39·9–44·3° (mechanical dryer, RH: 32–39%)	9·7		67·7			
Mackerel	*Rastrelliger kanagurta*	24 h in 30% salt (by weight) and saturated brine, dried with solar dryer for 48 h	7·28	3·39	46·55	19·19	15·35	0·89
Sardines and herrings	*Clupea fimbriata* *C. brachysoma* *C. perforata*	Combination of dry and wet salting (covered with 25% salt and dipped in 70% brine for 12 h). Dried with mechanical kiln 12 h at 40°C	9·09–12·5	2·41–4·19	43·16–47·70	31·15–37·34	15·69–18·20	0·85–0·82
Scads	*Decapterus* sp.	Dry salting, 25% (w/w) salt, for 68 h:						
		– sun-drying, 2 days	15·2	3·7	44·7	33·1	4·9	0·77
		– mechanical kiln (RH: 47–62·5%, 15 h/day, at 39·5–42·5°)	14·8	3·6	42·7	33·5	6·4	0·76
Shark	*Carcharhinus limbatus*	50 h in 30% salt and dried for 12 h at 40°C (mechanical kiln)	12·4		47·3	26·20	8·70	0·82
Silverbelly	*Leiognathus* sp.	25% salt for 37 h salting, 11·5 h drying at 39·9–44·3° (mechanical dryer, RH: 32–39%)	10·4		68·8			0·90

Dried								
Anchovy	*Stolephorus* sp.	Sun-dried, 15–18 h, 30–39°			20·06		10·02	0·90
		Solar dryer, 12–15 h, 36–58°			19·23		10·8	0·88
Smoked								
Catfish	*Cryptopterus* sp.	36 h smoking at 80–100°		3·96	16·10	76·70	3·24	0·83
Shark	*Carcharhinus limbatus*	Dipped in 10% brine, 30 min – 40–50°, 13 h – 70–80°, 13 h	1·4	1·7	38·9	30·1	4·0	0·98
Snakehead	*Ophiocephalus micropeltes*	36 h smoking at 80–100°		4·28	24·00	66·87	3·54	0·96
	O. striatus	36 h smoking at 80–100°		3·32	18·75	75·50	3·75	0·85
Smoked grilled								
Skipjack	*Katsuwonus pelamis*	– 2·5 h at 80–100°, smoke obtained from coconut shell			33·35	26·72		
		– 4 h at 80–100°, smoke obtained from mangrove wood			53·75	36·00		

TABLE 3

JAPAN: SMOKED AND DRIED FISHERY PRODUCTS

Product	*Species*	*Processing regime*	*Smoking and heating regimes*	*Composition (%)*			*Aw*[a]
				Fat	*Water*	*Ash*	
Plain dried fish							
Anchovy (Tazukuri)	*Anchovia indicus*	Whole body. Wash in fresh water	3–4 days by air-drying	5·0	16·0	11·8	0·79
Cod (Bo-tara)	*Gadus macrocephalus*	Dressed	Wind and air-drying	1·5	34·0	4·6	0·98
Herring (Migaki-nishin)	*Clupea pallasi*	Back meat fillet	2–3 weeks by air-drying	29·2	37·0	4·4	0·98
Sardine	*Sardinia melanosticta* *Etrumeus micropus* *Stelephores japonicus*	Round	Air drying (semi-dried)	15·6	59·5	2·8	0·98
Shark fin (White) (Black)	*Carcharhinus gangeticus* *Cynias manazo* *Isuropis glauca* *Prionace glauca* *Heterodontus japonicus*	Wash in sea water	2–3 weeks by air-drying	0	15·6	1·0	0·76
Squid (Surume)	*Todarodes pacificus* *Symplectoteuthis ovalaniencis*	Cut belly of mantle-cartilage. Wash with fresh water, 2–3° salt water, then fresh water	5–7 days at < 30°C sun-drying or dryer	4·4	18–20	6·6	0·84–0·88

	Watasenis scintillans					
	Sepiella maindroni					
	Sepia subaculeata					
	Sepia esculenta					
	Sepioteuthis lessoniana					
	Loligo edulis					
	Doryteuthis bleekeri					
Plain dried seaweed						
Laminaria (Konbu)	*Laminaria japonica*	1–2 days at < 30°C (sun-drying)	1·2	9·5	19·6	0·49
	L. ochctensis		2·0	10·0	20·3	0·53
	L. longipes					
	L. religiosa		0·4	10·3	22·1	0·54
	L. angustata		1·9	11·6	16·5	0·62
	L. coriacea					
	L. longipedalis					
	L. fragilis					
	L. diabolica					
	L. cichoriaides					
	Kjellmaniella crassifolia		0·5	12·3	24·7	0·65
	K. gyrata					
	Arthrothammus bifidus					
Purple laver (Nori)	*Porphyra tenera*	Sun-drying	1·9	11·1	6·9	0·60
	P. yezoensis					
	Enteromorpha sp.		0·3	7·3	14·1	0·35
	Monostroma sp.		1·0	16·9	12·4	0·80

(continued)

TABLE 3—*contd.*

Product	*Species*	*Processing regime*	*Smoking and heating regimes*	*Composition (%)*			*Aw*[a]
				Fat	*Water*	*Ash*	
Undaria (Wakame)	*Undaria pinnatifida* *Alaria crassifolia*	Mix with incompletely burned ash, then wash with sea water	Sun-drying	3·2	13·0	30·8	0·69
Boiled-dried							
Abalone (Hoshi-awabi) (Meibo) (Kaibo)	*Haliotis gigantea* *H. sieboldi* *H. discus hannai* *H. discus-discus*	Cure in 10% brine for 2–3 days. Boil at 50°C for 1–2 h	Repeat roasting (70°C) and drying (sun-dry) for 5–10 days	2·0	20·2	10·4	
Cartilage (Meikotsu)	*Mola mola* *Raja* sp.	Boil in hot water	Dry in sun				
Sand lance (Niboshi-ikanago)	*Ammodytes personatus*	Whole body. Boil in 10–12% brine for 10 min	2–3 days' sun-drying	5·0	46·2	14·8	0·98
Sardines (Niboshi-iwashi)	*Sardinia melanosticta* *Anchovia indicus*	Boil in 10–12% brine for 15 min	2–3 days' sun-drying	2·9	16·5	11·3	0·80
Sea-cucumber (Iriko)	*Stichopus japonicus* *Holothuria argus* *Pstsdyivhopud nigripunctatus* *Cucumaria japonica*	Boil in Bé 3° brine for 1–1·5 h	Roast at 70°C and dry in shade for 5 days				
Shellfish ligament	*Pecten yessoensis* *Pecten láqueatus* *Pinna pectinata japonica*	Boil in 3% brine for 3 min (first boiling). Boil in saturated brine for 10 min (second boiling)	Dry in sun or dryer at < 30°C for 4 weeks	2·6	17·8	8·3	

Shrimp Peeled after drying (Suri-ebi)	*Metapenaeus joyneri* *Palaemon nipponensis* *Sergestes lucens* *Pasiphaea sivado*	Boil in 2% brine for 30–40 min	Dry in dryer	4·0	19·5	11·6	
Peeled (Muki-ebi)	*Penaeus japonicus*	Boil shelled meat in 2% brine	Dry in dryer	2·8	35·2	15·0	
With shell (Kawatsuki-ebi)	*Metapenaeus joyneri* *Penaeus japonicus* *Palaemon nipponensis*	Whole body	Dry in dryer	3·0	21·6	12·3	0·93
Broiled-dried							
Flat fish	*Limanda herzensteini*	Whole fish	Broil on charcoal fire. Dry in sun	3·3	67·9	2·0	0·98
Sardine	*Sardinopus melanosticta*	Whole fish	Broil on charcoal fire. Dry in sun	19·9	51·0	5·8	0·98
Sea bream	*Pagrosomus major*	Eviscerated fish. Pierced with bamboo stick	Broil on charcoal fire. Dry in dryer	5·4	69·0	1·6	0·98
Dried-salted							
Alaska pollack fillet (Sukimi)	*Theragra chalcogramma*	Fillet without skin. Dry salt	Dry in air	0·9	35·7	23·3	
Cod	*Gadus macrocephalus*	Split at dorsal side. Cure 2 days in Bé 18° brine. Press with weight for 1 day	Dry in air	1·5	34·0	4·6	
Jack mackerel (Shioboshi-aji)	*Trachurus japonicus*	1 night with dry salt	Dry in air	6·8	68·1	4·8	

(continued)

TABLE 3—*contd.*

Product	*Species*	*Processing regime*	*Smoking and heating regimes*	*Composition (%)*			*Aw*[a]
				Fat	*Water*	*Ash*	
Mullet roe (Karasumi)	*Mugil cephalus*	Unbroken mullet roe. Dry salt with 12% (w/w) for 10–12 h	Dry in sun for 20 days	26·0	23·0	6·4	
Sardine Round	*Sardinopus melonosticta*	6–8 h in Bé 15° brine	Dry in air	23·2	30·5	10·5	
	Etrumeus micropus			4.8	36·1	11·7	
Split	*Sardinopus melanosticta*	1 night in Bé 15°–18° brine	Dry in air	15·6	59·5	2·8	
Skewered	*Sardinopus melanosticta*	4–5 h in Bé 6°–7° brine	Dry in air for 5–6 days	21·9	54·5	4·6	
Scad (Kusaya-no-himono)	*Decapterus muroadsi*	1 night in repeatedly used brine	Dry in sun	3·0	38·6	8·2	
Yellowtail (Enkanburi)	*Seriola dorsalis*	Split ventral side. 4–5 days with dry salt	Dry in sun				
Bound with straw (Waramaki-buri)	*Seriola dorsalis*	Split ventral side. 1 week with dry salt	Dry in shade for 1–2 weeks				
Small	*Seriola dorsalis*	Fillet without skin. 1–1½ days with dry salt	Dry in sun for 7–10 days				
Frozen-dried							
Agar	*Gelidium amansii*	Extract gel by heating seaweed raw material	Repeat freezing and drying	0	19·9	3·2	0·86
Alaska pollack (Mintai)	*Theragra chalcogramma*	Place fish outdoors	Repeat freezing and drying for 70 days	1·5	34·0	4·6	0·98
Dried Fish Sticks							
Bonito (Katsuobushi)	*Gymnosarda epelamis* *Sarda chilensis*	Fillet without skin. Boil for 60–90 min. Mould 3 times	Broil in smoking furnace. Repeat sun-drying for 1 month	2·9	15·2	4·0	0·76

Mackerel (Saba-bushi)	*Scomber japonicus*	Fillet with skin. Boil for 30–60 min	Broil in smoking kiln. Dry in dryer	8·0	9·5	3·0	0·54
Smoked							
Herring Cold smoked (Reikun-nishin)	*Clupea pallasi*	Round. Dry salt with 12–20% (w/w) for 1 week. Desalt with fresh water	Smoke-dry for 3–4 weeks	31·1	37·8	7·9	
Hot smoked (Onkun-nishin)	*Clupea pallasi*	Cut back portion. Cure in 10–15% brine for 40–90 min	Smoke-dry for 4–6 h from 20°C to 85°C				
Salmon Cold smoked (Reikun-sake)	*Oncorhynchus tschawytscha* *O. nerka* *O. keta*	Dry salt for 10–30 days. Desalt 1–3 days with fresh water	Smoke for first week at 18°C, second week at 22°C and third week at 25°C	7·3	48·9	10·0	
Hot smoked (Onkun-sake)	*Oncorhynchus nerka*	Fillet with skin. Cure in 15–18% brine for 5–6 h	Smoke-dry for 7 h at 50°C, 2–3 h at 80°C	5·6	60·0	8·0	
Yellowtail Cold smoked	*Seriola quinqueradiata*	Fillet with skin. Cure in saturated brine containing 5% sugar and small amount of spice for 8–15 days. Desalt for 2–3 days in fresh water	Smoke-dry for first 4–5 days at 20–25°C, second 3 days at 25–30°C, and third 3 days at 30–35°C				
Seasoned-smoked							
Squid (Ika-kun)	*Ommastrephes bartrami*	Dress and skin. Boil at 80–90°C for 2–3 min. Cure in 5–10% brine with 30–40% sugar for 1 night	Smoke-dry for 1–2 h at 20–25°C, 2–3 h at 50–70°C	2·5	38·5	6·2	0·98

[a] A salt content of 1·5% has been assumed for unbrined fish where measured salt content is not reported.

TABLE 4

MALAYSIA: DRIED FISHERY PRODUCTS

Product	Species	Preparation and salting	Drying	Composition (%, wet weight basis)			Aw
				Salt	Fat	Water	
Anchovy (Bilis)	*Stolephorus commersonii*	Whole fish boiled in 10% brine in bamboo baskets for 1 min with agitation during cooking on board boat. Baskets removed from brine and allowed to drain and cool (West Coast)	Fish spread to dry in the sun for 1 day	HG: 9·2 LG: 12·0	HG: 3·0 LG: 1·7	HG: 29·5 LG: 38·9	0·79 0·79
Catfish (Duri)	*Tachysurus caelatus*	Fish split, gutted and snout cut off. Put in concrete vat with pickle from previous salting batches for 1 day (East Coast)	Sun-drying	12·7	4·3	47·5	0·82
Jewfish (Gelama)	*Johnius (Pseudosciaena) soldado*	Whole and split fish pickle salted in the same way as mackerel	Sun-dried as for queenfish but for $1\frac{1}{2}$–2 days	W: [a]19·5, 15·3 S: 16·7	W: 3·5 ND S: 0·6	W: [a]42·5, 48·7 S: 43·7	0·75

Mackerel (Kembong)	*Rastrelliger* sp.	Whole fish pickle salted for 2 days in same way as queenfish but with less salt, 100 parts fish:20 parts salt	Sun-dried as for queenfish but for only one day	16·1	6·4	48·6	0·77
Octopus (Sotong)	Unknown	Octopuses are slit open, gutted, washed and left in coarse salt for 3 h	Sun-dried on racks (which are either laid horizontally or slanted and rotated to follow the sun during drying) for several days until they become translucent	6·9	0·6	33·6	0·86
Pomfret (Bawal)	*Pampus argenteus*	Split fish pickle salted as for mackerel	Sun-dried as for queenfish but for shorter time	[a]20·0, 13·9	1·1 ND	[a]55·6, 48·0	0·75
Queenfish (Talang)	*Scomberoides* sp.	Fish are scaled, split in butterfly cut, gutted (gills left intact) and washed. Pickle salted for 2 days using 100 parts fish:30 parts salt. Dry salt first rubbed in flesh; then fish packed in dry salt in wooden tubs for one day.	Washed in freshwater then dried on rattan mats on the ground for 2–5 days	[a]14·2, 14·6	[a]0·9 ND	[a]53·5, 53·7	0·82

(continued)

TABLE 4—*contd.*

Product	*Species*	*Preparation and salting*	*Drying*	*Composition (%, wet weight basis)*			*Aw*
				Salt	*Fat*	*Water*	
		On second day, pickle is added to cover fish which are weighted down during salting					
Red snapper (Merah)	*Lutianus argenti-maculatus*	As for queenfish	As for queenfish	[a]15·1, 14·5	[a]0·3 ND	[a]51·6, 51·3	0·80
Sardine (Tamban)	*Sardinella* sp.	As for anchovy	As for anchovy but sun-dried for only 4 h	7·5	3·6	22·5	0·75
Shrimp (Udang) imported from Thailand	Unknown	Boiled in brine for a very short time	Sun-dried on mats on the ground for about 1 day	HG: 10·2 LG: 13·3	HG: 1·9 LG: 1·0	HG: 32·4 LG: 32·9	0·78 0·75
Squid (Sotong)	Unknown	Squid are slit and the quill (back-bone), ink sac and guts removed. No salting	Sun-dried on racks after washing	2·8	3·2	26·3	0·93

HG: High grade.
LG: Low grade.
W: Whole.
S: Split.
ND: Not determined for second batch.
[a] Two separate batches of fish purchased at different times.
NB There are regional differences in processing.

TABLE 5

NORWAY: SMOKED AND DRIED FISHERY PRODUCTS

Product	*Species*	*Processing/production*	*Composition (%)*			*Aw*
			Salt	*Fat*	*Water*	
Smoked						
Greenland halibut Fillet of fresh Greenland halibut with skin and pin bones, or raw material that has been frozen while fresh, and then thawed	*Reinhardtius hippoglossoides*	Remove loose scales by washing. Sugar-cure in tight vessel. Duration of curing: 4–6 days, depending on thickness of fish. After curing, the fish is to be hung in fresh water to soak for 8–12 h until the desired salt content has been achieved, and then dried for 12 h at room temperature Cold smoking in chamber kiln for 20 h. In mechanical kilns with great smoke density, the smoking period may be reduced For long-term storage, smoked Greenland halibut must be vacuum packed and stored frozen in air-tight wrappings	1–3	18–25	50–60	0·98
Haddock — lightly smoked fillets Fillet of fresh haddock with skin and pin bones. Raw materials that have been frozen and thawed will result in a less juicy product	*Melanogrammus aeglefinus*	The fillets are washed to remove blood, spots and loose scales. Soak in 20 % brine for 10–20 min depending on size of fish. Then dry on rack with skin-side down for 2 h				

TABLE 5—*contd.*

Product	*Species*	*Processing/production*	*Composition (%)*			*Aw*
			Salt	*Fat*	*Water*	
		Smoke-curing in mechanical kiln for 2–3 h at a max. temperature of +30°C. After smoking, cool at air temperature for at least 2 h. The fillets may be frozen individually and glazed for long-term storage				
Haddock — split Fresh haddock without taste of sea bottom or other after-taste. Frozen and defrosted raw materials result in a dry product	*Melanogrammus aeglefinus*	The fish is skinned and black membranes removed. Remains of blood and loose scales are removed. The fish is salted in containers with closed bungs (tight vessels) with 1 hl salt per 1000 kg skinned fish Duration of salting period: 15–20 h according to size of fish Drainage and pre-drying for 20 h in air with relative humidity of 65–70 % Cold smoking in chamber kiln for 20 h. In kilns with greater smoke density, the period of smoking may be reduced	2–5	0·5	70–75	0·98

Herring — Buckling Fresh, round herring, nobbed or gutted, often with hard or soft roe left in fish	*Clupea harengus*	Dry salting or salting in saturated brine for 2–3 h. Nobbed herring is cured in 70–80 % brine for 1 h. Herring is hung in kiln while wet with brine and smoked at 27–32°C for 1 h and then at 71–77°C for 1 h. Herring with head and guts should be smoked with a final temperature of 95°C. Store in cool storage facilities	1–3	10–20	55–65	0·98
Herring — Kippers Fresh or frozen herring of good quality fat herring or completely fresh, spawning large herring	*Clupea harengus*	Wash and remove loose scales from herring. Split along back so that backbone remains on left side of fish. The head is not removed, but guts and roe are taken away. Pre-marination in 80 % brine for 15 min Hang for drainage on net racks or hooks for about 1 h. To be smoked at max. 29°C in mechanical kiln for 4 h. Smoke temperature max. +35°C In traditional kiln max. +30°C for 12–15 h. Will keep for 3 weeks in cool storage	1–3	13–15	65	0·98
Mackerel Fully salted, autumn mackerel cleaned, with head	*Scomber scombrus*	Soak well in running water for 3–4 days. Dry on net racks in frames at room temperature for 15–20 h. Cold smoking in chamber kiln for 24 h. Curing time may be reduced in mechanical kiln	2–5	*c.* 11	*c.* 65	0·95–0·98

(*continued*)

TABLE 5—*contd.*

Product	*Species*	*Processing/production*	*Composition (%)*			*Aw*
			Salt	*Fat*	*Water*	
		Smoked mackerel can be kept in cold storage for about 1 week When storing for longer periods, the fish must be frozen in air-tight wrappings				
Saithe as salmon (salmon substitute) Fully salted, split saithe or saithe fillets of fresh, salted raw materials	*Pollachius virens*	Soak the salted fish: during the summer in a 15% salt solution for 24 h; during the winter in a 10–12% salt solution for 24 h. Generally speaking, the temperature of the brine (salt solution) determines the length of time the fish should be soaked. Slice fillets by hand or machine. Lay slices in thin layers in basket and dip in dye. Dye period: 12 min. Cold smoking at a max. temperature of 30°C for 24 min. Pack slices in air-tight wrapping and add Norwegian oil or olive oil before sealing wrappings The length of time the product will keep depends on salt content and water content. In air-tight wrappings, the product must be stored in cold-storage facilities	8–10	3–10	60–63	0·88–0·91

Salmon Completely fresh salmon with a fat content of up to 15%. Salmon that has been frozen and thawed will result in a dry product	*Salmo salar*	The fish is to be split along the back. Head, backbone and tail fin to be removed. The side fillets with skin and shoulder bone are joined across the belly. The fish is washed in ice-cold water, drained and dry-salted in relatively finely ground sea salt, with a mixture of 2 parts salt to 1 part sugar. Cover fish well with salt mixture Salting period: 2–5 days, depending on size of fish After salting, the salt is rinsed off the fish, and it is spread on wooden stays and hung up to air-dry for 10–12 h. Cold smoking in chamber kiln for 20–36 h depending on size of fish When curing salmon fillets in 70–80% brine, curing time is 3–4 h Cold smoking in mechanical kiln should last for 5–12 h depending on thickness of fish. Smoked salmon should be kept in cold storage. When storing for long periods, salmon may be frozen in air-tight wrappings	2–5	*c.* 8	*c.* 65	0·95–0·98
Salted and dried fish						
Cod — split Fresh fish bled while alive. The raw material to be washed and loose slime removed	*Gadus morhua*	The fish is split: opened along belly so its remains joined at back. Backbone is removed from the third vertebra behind anal opening and towards head	Lightly salted: 4–5		70	0·98

(continued)

TABLE 5—*contd.*

Product	*Species*	*Processing/production*	*Composition (%)*			*Aw*
			Salt	*Fat*	*Water*	
		Clean fish for remaining blood and intestines, and wash well in clean running water Dry-salt in layers with 30–33 kg fish-salt to 100 kg fish After 4–5 days, the fish is repacked and resalted with 18–20 kg salt per 100 kg fish Fish to remain in salt for a total of 3 weeks (ripe in salt)	Ripe in salt: 18–20		56–57	0·77
Cod — Finnmarken Round cod Quality requirements as for Lofoten round cod, but the fish is slightly smaller	*Gadus morhua*	Regulation cleaning and head removal. Air-drying in fresh air for 2–2½ months at an air temperature of no less than 0°C. Ideal temperatures: +8–+10°C. Relative humidity 65–75% The product: a differentiation is made between prime quality, second quality and Africa cod which is sorted into several sizes		1·3	15–16	0·75
Cod — Klippfish 7/8 dry Fully salted cod (ripe in salt). Quality of fresh fish before salting as for Salted fish – Cod, split	*Gadus morhua*	Cod ripe in salt to be washed in running water in special washing installation or by hand. Fish to be stacked to press out water. In the case of traditional drying, the fish is to be dried on a rack according	18–20		45	0·75

		to a pre-determined drying programme with three drying periods and restacking between each period Temperature: 16–18°C at first stage, and max 20°C at last stage Relative Humidity: 30–45% at beginning of drying periods, and 60–80% at end Max. speed of drying air: 2m/s Drying time for traditional drying with restackings: 2 weeks Drying in installations with dehydration of drying air and exact automatic adjustment of temperatures and humidity, without restacking: drying time 2–3 days Quality: extra, superior, universal, popular			
Cod — Stockfish Lofoten Round Cod Fresh line-caught fish generally gives the best quality. Fish that has remained too long in the nets gives second and third grade fish. The minimum size of the fish may vary from year to year	*Gadus morhua*	Regulation cleaning. When removing head, at least one vertebra shall accompany the head. The fish is to be dried on flakes or drying frames. Slow drying in fresh air for 2–3 months at +5–+10°C. Relative humidity 65–75% The product: a differentiation is made between prime quality fish of several varieties and sizes, and second quality fish in three sizes and Africa round cod	1·3 (for prime quality fish)	15–16	0·75

(continued)

TABLE 5—*contd.*

Product	*Species*	*Processing/production*	*Composition (%)*			*Aw*
			Salt	*Fat*	*Water*	
Cod — Stockfish Split cod Requirements to raw materials as for round cod. Prime quality requires muscular and well-rounded fish for raw material	*Gadus morhua*	The fish is split lengthwise, but remains joined at the tail. About 2/3 of the backbone from the neck and back is removed. The lower part of the backbone remains in the one half of the fish. The split is then tied around the tail with sisal or hemp string. After splitting and cleaning, the fish is rinsed carefully in fresh water, and is then hung to dry, preferably on a wire-strung rack. Split dries somewhat more quickly than roundfish ($1\frac{1}{2}$–2 months). Relative humidity 70–80 %. Post-drying in well-aired storerooms	1·8	1·3	15	0·74
Haddock — Stockfish Roundfish Requirements to raw materials as round cod. The scales of prime quality fish are not to be removed	*Melanogrammus aeglefinus*	Regulation cleaning. When removing head, at least one vertebra shall accompany the head. The fish is to be dried on flakes or drying racks Slow drying of fish in fresh air for 2 months Relative humidity in air: 65–75 %		1·4	15–16	0·74

Herring — cured with salt and spices. Fat herring, fished in autumn or early winter, head removed and belly-cleaned	*Clupea harengus*	Fish with head removed to be carefully washed to remove loose scales. Salted in barrels. 13–15 kg fine salt + 6 kg sugar + 1·0–1·4 kg spice mixture per 110 kg herring. Refill with saturated brine 4–8 days after spicing. Mature after 2 months. To be stored at +2°C. Will keep 5–6 months
Herring — Heavy salted (fully salted herring). Fat herring, freshly gutted, rarely belly-cleaned. Herring fished in the autumn results in the best quality	*Clupea harengus*	Salted in barrels. 28–35 kg salt per 110 kg fresh herring. Relatively finely ground sea salt. Matured after 3–4 weeks. Temperatures: +10–+12°C. After maturing, barrels to be refilled with saturated brine. Will keep for 1 year at +10°C
Herring — Sugar–salted. Fat herring, freshly gutted and belly-cleaned, or with head removed and belly-cleaned	*Clupea harengus*	Salted in barrels. 15 kg salt + 6 kg sugar per 110 kg herring. To be refilled with saturated brine. Matured after 3–4 weeks at +10°C. Will keep for 8 months at +10°C
Herring — Matjes type. Fat herring with a slightly developed hard and soft roe, gutted and belly-cleaned or with head removed and belly-cleaned	*Clupea harengus*	Immediately after gutting, the herring is salted in barrels. 15–16 kg relatively finely ground sea salt per 100 kg fresh herring. Strength of brine in barrel: 18–22°Bé. Refill with brine, strength 25°Bé

TABLE 6

PHILIPPINES: DRIED FISHERY PRODUCTS

Product	*Species*	*Salting regime*	*Salting/ drying regime*	*Composition (%)*					*Aw*
				Salt	*Fat*	*Water*	*Protein*	*Ash*	
Dried									
Frigate mackerel	*Auxis thazard* (Lacépède)	21 % brine	9·5 h	6·67	2·50	30·30	40·60	18·20	0·85
Mussel	*Mytilus smaragdinus*	3 % brine	5·5 h	1·46	–	24·00	33·66	–	0·95
Oyster	*Ostrea* spp.	3 % brine	5·5 h	10·85	1·19	23·00	31·75	15·58	0·73
Roundscad	*Decapterus macrosoma*	21 % brine	9 h	12·00	9·09	31·10	58·00	15·80	0·75
Squid	*Loligo* spp.	–	8 h	1·84	4·40	25·40	53·83	5·59	0·97

TABLE 7

POLAND: SMOKED FISHERY PRODUCTS

Product	*Species*	*Brining regime*	*Drying, heating and smoking regime*			*Loss during smoking (%)*	*Salt content after salting (%)*
				°C	*min*		
Hot smoked							
Alaska pollock	*Theragra*	25–45 min	I[a]	65–85	60–80	42–45	–
gutted without head	*chalcogramma*	20°Bé	II	85–115	50–60		
			III	90–100	50–65		
Bream	Sparidae	20–40 min	I	45–70	30–70	43–47	1·5–2·0
gutted without head		20°Bé	II	90–115	40–60		
			III	80–100	30–70		
chunks		35–50 min	I	60–90	40–80	32–37	1·5–2·0
		20°Bé	II	90–120	50–70		
			III	80–100	60–90		
Carp	*Cyprinus carpio*	30–60 min	I	45–60	45–60	35–42	1·2–2·0
gutted		20°Bé	II	90–110	40–50		
			III	60–80	50–60		
Cod	*Gadus morhua*	60–90 min	I	50–90	40–60	38–42	1·2–1·8
gutted without head		20°Bé	II	90–120	40–60		
			III	90–110	60–90		
Cutlass fish	*Trichiurus* sp.	15–20 min	I	45–60	50–60	32–36	–
chunks, gutted		20°Bé	II	90–110	45–60		
			III	70–80	70–80		
Dentex	*Dentex dentex*	35–50 min	I	60–90	40–80	–	1·5–2·0
chunks		20°Bé	II	90–120	50–70		
			III	80–100	60–90		

(*continued*)

TABLE 7—*contd.*

Product	*Species*	*Brining regime*	*Drying, heating and smoking regime*			*Loss during smoking (%)*	*Salt content after salting (%)*
				°C	*min*		
Eel gutted	*Anguilla* sp.	60–90 min 20°Bé	I	45–60	40–60	20–28	1·5–2·2
			II	100–120	35–60		
			III	80–100	45–60		
Flounder gutted	*Pleuronectidae*	40–80 min 20°Bé	I	30–60	40–90	43–47	1·2–2·5
			II	90–115	40–60		
			III	90–100	50–60		
Gilt sardine	*Sardinella aurita*	90–130 min 20°Bé	I	40–60	90–120	27–34	1·0–2·0
			II	90–120	60–70		
			III	80–110	70–90		
Hake gutted without head	*Merluccius merluccius*	60–90 min 20°Bé	I	45–90	40–60	35–43	1·2–1·8
			II	90–120	40–60		
			III	90–110	60–90		
Halibut chunks	*Hippoglossus hippoglossus*	30–40 min 20°Bé	I	30–90	50–70	44–49	1·0–1·5
			II	90–110	30–60		
			III	70–100	50–70		
Herring buckling	*Clupea harengus*	60–90 min 20°Bé	I	40–80	90–100	29–35	1·2–2·0
			II	80–110	50–60		
			III	70–100	50–70		
Boston	*Alosa alosa*	30–60 min 20°Bé	I	40–60	90–100	31–35	1·2–2·0
			II	90–110	50–60		
			III	70–100	70–80		
gutted	*Clupea harengus*	20–40 min 20°Bé	I	40–60	60–80	–	1·2–2·0
			II	80–100	50–60		
			III	70–90	40–60		
Horse mackerel	*Trachurus* sp.	35–50 min	I	50–80	40–70	26–31	1·2–2·0
			II	90–115	50–60		
			III	80–100	40–70		

gutted with or without head	*Trachurus* sp.	35–50 min	I	50–80	40–70	26–32	1·2–2·0
			II	90–115	50–60		
			III	80–100	40–70		
Mackerel	*Scomber scombrus*	20–40 min	I	70–80	40–60	23–27	1·5–2·5
		20°Bé	II	85–110	30–40		
			III	90–100	40–60		
fillets with garlic	*Scomber scombrus*	60–80 min	I	40–60	20–30	34–38	1·2–2·0
		14°Bé with	II	90–100	30–40		
		garlic	III	70–90	50–80		
gutted without head	*Scomber scombrus*	20–40 min	I	70–85	40–60	25–33	1·5–2·0
		20°Bé	II	85–110	30–40		
			III	90–100	40–50		
Salmon chunks	*Salmo salar*	15–25 min	I	40–60	45–60	–	–
			II	80–110	40–50		
			III	50–90	40–50		
Sprat	*Sprattus sprattus*	10–20 min	I	50–60	10–20	40–45	1·0–1·7
		15°Bé	II	80–110	10–20		
			III	60–90	30–40		
Cold smoked							
Atlantic bonito fillets	*Sarda sarda*	3–4 days in salt + sugar, + spices		40	15–30	–	–
Halibut chunks	*Hippoglossus hippoglossus*	5–8 days in salt + sugar,	I	20–25	3–6	–	3–5
		12–18°C	III	50	40–70		
Herring fillets	*Clupea harengus*	20 min 20°Bé		20–40	12–15	29–35	–
Horse mackerel à la kippers	*Trachurus* sp.	50 min 20°Bé		20–30	12–15	–	–
Mackerel fillets	*Scomber scombrus*	20–50 min 18 %		20–40	12–15	32–37	4–5
Salmon sides	*Salmo salar*	4–6 days in salt + sugar,	I	25	3–6	–	4–6
		+ $NaNO_3$ 15–18°C	III	30	40–70		

[a] I – Pre-drying; II – heat processing; III – smoking in heavy smoke.

TABLE 8

SRI LANKA: SMOKED AND DRIED FISHERY PRODUCTS

Product	*Species*	*Salting regime*	*Smoking and/or heating regime*	*Composition (%)*				*Aw*
				Salt	*Fat*	*Water*	*Protein*	
Smoked Fish								
Herring hot smoked	*Amblygaster sirm*	25% salt for 1–2 h	4–6 h at 60–90°C in a mechanical kiln	1·5–2	3–5	50–55	30–35	0·98
cold smoked (kipper)	*Amblygaster sirm*	2–3 h in 25% salt	4–6 h at 35–40°C in a kiln	8–10	4–5	55–60	26–38	0·90–0·88
Skipjack	*Katsuwonus pelamis*	Dry salted for $\frac{1}{2}$ h or cooked in brine for a few minutes	Smoked and dried on coals for 10–12 h *or* Smoked and dried on paddy husk for 6–10 h *or* Sun-dried for 6–7 days	4–6	3–5	14–17	72–80	0·64–0·67
Yellow fin tuna salted and hot smoked (Maldive fish)	*Thunnus albacore*							

Dried Fish								
Herring	*Amblygaster sirm*	Saturated brine 12–24 h	Sun-dried on sandy beaches or mats for 3–4 days 6–8 h/day	15–18	4–5	35–40	35–40	0·75
Leather skin	*Chorinemus lysan*	Dry salted	Sun-dried on sandy beaches or mats for 4–6 days	12–15	1–1·5	40–45	40–45	0·79–0·77
Sea pike	*Sphyraena* sp.	Dry salted	Sun-dried on sandy beaches or mats for 4–6 days	14–15	3·5–4	40–45	35–40	0·76–0·77
Silverbelly	*Leiognathus* sp.	Saturated brine 6–12 h	Sun-dried on sandy beaches or mats for 4–6 days	10–15	3–5	30–35	35–40	0·77–0·75
Skipjack	*Katsuwonus pelamis*	Dry salted	Sun-dried on sandy beaches or mats for 4–6 days	10–12	1–3	40–45	40–45	0·83–0·82
Tilapia	*Tilapia mossambica*	Dry salted	Sun-dried for 4–8 days	10–15	3–4	30–35	35–40	0·77–0·75

TABLE 9

THAILAND: SMOKED AND DRIED FISHERY PRODUCTS

Product	*Species*	*Salting regime*	*Smoking and drying regime*	*Composition (%)*				*Aw*
				Salt	*Fat*	*Water*	*Pro-tein*	
Dried salted								
Anchovy	*Anchoviella commersonii*	10 % brining for ½–1 h	Washed with water before drying	12·44		58·4	29·2	0·65
Anchovy, silverside	*Stolephorus* sp.	10 % brining for 15–20 min	Washed with water before drying		4·0	13·4	68·2	
Anchovy, silverside (boiled)		Boiled in saturated brine for few min	Sun-dried		1·6	44·8	37·8	
Bream, freshwater	*Sparus burda*	Dry salted, cured for 1–2 days	Sun-dried		6·2	43·8	34·3	
Croaker	*Pseudosciaena* sp.	Dry salted, cured for 1–2 days	Sun-dried		5·8	14·5	46·2	0·82
Mackerel	*Rastrelliger brachysoma*	3 days in 1:3 ratio of salt to fish (dry salting)	Washed with water before drying	13·69		44·62	34·7	0·73
Mackerel, rake-gilled; Mackerel, striped	*Rastrelliger* sp.	3 days in 1:3 ratio of salt to fish (dry salting)	Washed with water before drying		5·2	37·7	42·0	
Mackerel, short bodied	*R. neglectus*	3 days in 1:3 ratio of salt to fish (dry salting)	Washed with water before drying		8·6	47·3	28·6	

Mackerel, Spanish	*Scomberomorus commersonii*	Saturated brining for 8 days	Washed with water before drying		4·0	45·8	35·1	
Nemipterid, ribbon finned (semi-dried)	*Nemipterus taeniterus*	Dry salted, cured for 1–2 days	Sun-dried for 6–12 h		2·9	43·9	34·6	
Queenfish, slender	*Scomberoides tol.*	Dry salted, cured for 1–2 days	Washed with water before drying		2·5	42·8	35·3	
Sardine, fringescale	*Sardinella fimbriata*	Saturated brining	Sun-dried		2·4	45·8	38·0	
Sepat Siam	*Trichogaster pectoralis*	1 day in 1:9 ratio of salt to fish (dry salting)	Sun-dried		9·0	36·2	45·3	0·84
				7·16		43·3	48·36	
Shad, gizzard	*Anodontostoma chacunda*	Dry salted, cured for 1–2 days	Washed with water before drying		5·4	51·5	23·7	
Shark, fins	*Carcharias* sp. *Scoliodon* sp.	Saturated brining for 30 min	Sun-dried		0·3	12·8	84·9	
Skipjack, strips	*Katsuwonus pelamis*	Saturated brining	Washed with water before drying		5·1	15·3	75·6	
Skipjack, strips (semi-dried)	*Katsuwonus pelamis*	Saturated brining	Sun-dried		6·2	39·6	51·5	
Snakehead, murrel	*Ophiocephalus striatus*	Dry salted, cured for 2–3 days	Washed with water before sun-drying		7·0	38·1	45·7	
				6·11		62·2	20·57	
Threadfin, spiny	*Polynemus* sp.	Dry salted, cured for 2–5 days	Washed with water before drying		1·8	46·4	40·2	
Prawn, river / Shrimp, common	*Leander* sp.	Boiled in 10% brine for 5 min	Sun-dried on cement ground or mats for 1–2 days		3·0	29·3	59·8	
Shrimp	*Leander* sp.	Boiled in 10% brine for 5 min	Sun-dried on cement ground or mats for 1–2 days	9–13		30–35		0·72–0·76

(continued)

TABLE 9—*contd.*

Product	Species	Salting regime	Smoking and drying regime	Composition (%) Salt	Fat	Water	Protein	Aw
Shrimp	*Palaemon* sp. *Penaeus* sp.	Boiled in 10% brine for 5 min	Sun-dried for 1–2 days		3·5	13·7	62·4	
Abalone, earshell	*Haliotis gigantea*				1·5	27·3	48·5	
Clam, bean	*Donax faba*	1 h in 1:10 ratio of salt to fish (dry salting)	Sun-dried		2·0	24·6	48·1	
Clam, razor	*Solenidae* sp.	1 h in 1:10 ratio of salt to fish (dry salting)	Sun-dried		2·9	25·0	43·4	
				11·33		31·58	18·81	
Cuttlefish	*Sepia* sp.	2·5% brining for 2–3 min	Sun-dried		4·4	18·1	67·5	0·66
Mussel, saltwater	*Mytilus samaragdinus*	Cooked whole shell for 10 min the body then boiled in brine for 5 min	Sun-dried for 2 days or 55–60°C in kiln for 2–4 h		4·9	25·0	42·9	0·80
				4·71		37·6	45·41	
Mussel, sea or horse	*Mytilus* sp. *Modiola* sp.	Dry salted in ratio of salt and body 1:10 (by weight), cured for 1 h	Sun-dried 1 day		7·2	13·8	51·9	
Oyster	*Ostrea* spp.	Dry salted in ratio of salt and body 1:10 (by weight), cured for 1 h	Sun-dried 1 day		5·1	18·8	35·4	

Squid	*Loligo* sp.	2·5% brining for 2–3 min	Sun-dried		4·6	21·8	63·3	0·78
Squid, sweet	*Loligo duvauchelli* / *Loligo formosana*	2·5% brining for 2–3 min	Sun-dried	14·5		38·2	44·68	0·60
Sea cucumber	*Holothuria* sp.	Boiled in water for 20–30 min	Sun-dried		1·7	8·9	82·0	
Smoked								
Mackerel, rake-gilled striped	*Rastrelliger* sp.	Saturated brining for 5 min	Sun-dried for 1 h then smoked and dried on paddy husk for 3 h at 50–55°C		1·7	62·1	30·4	
Milk-fish	*Chanos chanos*	2–6 h in 1:2 ratio of salt to fish (dry salting)	Sun-dried		8·3	60·5	29·3	
Mullet, harder; salted roe	*Mugil* sp.	Saturated brining for 1–2 h or 6–8 h	Smoked and dried at 40–60°C for 24–36 h or 40–70 h		26	23·0	40	
Hot smoked								
Crispy dried fish	*Kryptopterus limpok*	Gutted head-on fish is used. Very lightly salted. Cook over the open charcoal fire. When the outside of product is sufficiently dried, hot smoke is applied using sawdust to cover the red hot charcoal and the smoke is liberally generated. The smoke procedure is continued until the fish is completely dried						

TABLE 10

UNITED KINGDOM: SMOKED FISHERY PRODUCTS

Product	Species	Smoking regime[a]		Composition (%)[a]					Aw
		Temperature (°C)	Time (h)	Salt	Fat	Water	Protein	Ash	
Smoked fish									
Cod	*Gadus morhua*								
Skinned fillets		Cold	3	2–3	< 1	75–79	20–27	13–22	0·98
Roe		32–38	8	4–7	3	60			
Dogfish	*Galeorhinus* sp.								
Flaps	*Mustelus* sp. *Scyliorhinus* sp. *Galeus melastomus* *Squalus acanthias*	Hot 90°C from start	1		2	26–30	50	18·7	0·97
Eels	*Anguilla* sp.								
Small		Hot	$1\frac{1}{4}$–$2\frac{1}{2}$	4–6	30	40			0·90
Large		Hot	$2\frac{1}{4}$						
Haddock	*Melanogrammus aeglefinus*								
Skin on fillets		Cold	3	2	< 1	72	22	4	0·98
Finnans		Cold	4–6	2–3	< 1	75–79			0·98
Pale cure		Cold	2–3		< 1				
Cutlets: single block fillets		Cold	2–3	2–3	< 1	75–79			0·98
Arbroath smokies		Hot	2	2–3	< 1	70–75			0·98
Herring	*Clupea harengus*								
Kippered		Cold	3–4	1·5–2·5	10–20	45–72	21		0·98
Fillets (skin on)		Cold	2–3	1·5–2·5	10–20	45–72	21		
Kippered for canning		Cold	2–$2\frac{1}{2}$	2–3	15–20	45–55	21		
Canning		Hot	2	2–2·5	15–20	45–72			

Bloaters (whole fish)		35–40	3	2–3	15–20	45–55			
Buckling		Hot	3	2–3	15–20	45–55			
Red		Cold	6	14	15–20	40–45			
Mackerel	*Scomber scombrus*								
Kippered fillets		Cold	3	2–2·5	15–20	45–65	20–25	6	0·98
Fillets		Hot	$2\frac{1}{2}$–3	2–2·5	15–20	45–65	20–22	6	
Whole, gutted		Hot	3–$3\frac{1}{2}$	2–2·5	15–20	40–65	20–22	6	
Plaice	*Pleuronectes platessa*	Hot	2						
Redfish	*Sebastes* sp.								
Skinned fillets	*Helicolenus dactylopterus*	Cold	3	2		75–79			0·98
Steaks		Hot							
Saithe	*Pollachius virens*								
Skinned fillets		Cold	3	2–3		75–79			0·98
Salmon	*Salmo salar*								
Fillets		Cold	4–10	5	10–20	55–60	25–35	13·5	0·94
Sprat	*Sprattus sprattus*								
Canning		Hot (heavy smoke)	1–$1\frac{1}{2}$		30	50	18	2·6	0·98
Kielersprotten		Hot (light smoke)	1–$1\frac{1}{2}$						
Trout, rainbow	*Salmo gairdneri*	Hot	$2\frac{1}{2}$–3	2–6		65–72			0·93–0·98
Trout, sea (salmon)	*Salmo trutta*	Cold		3–5	8–15	55–65	25–30		0·94–0·98
Whiting	*Merlangius merlangus*								
Cutlets/skin off or on		Cold	2–3	2–3	< 1	75–79			0·98
Smoked shellfish									
Mussels	*Mytilus edulis*	Hot	$\frac{1}{2}$–$\frac{3}{4}$						
Oysters	*Crassostrea* sp.	Hot	$\frac{1}{2}$–$\frac{3}{4}$						

(continued)

For notes see page 156.

[a] NOTES

Cold smoking. Product temperature rarely exceeds 25°C for smoke temperature of 27°C to 30°C. In a mechanical, or Torry type, kiln the product temperature is close to the wet bulb temperature which is commonly 5°C below the dry bulb (smoke) temperature.

Hot smoking. Product temperature rarely exceeds 70°C for smoke temperatures up to 80°C. Depending on the thickness of the product the centre product temperature is commonly some 10°C below dry bulb temperature, but can range from 5°C to 20°C below the dry bulb temperature. Internal temperatures are unlikely to exceed 60°C for greater than 30 min in many UK hot smoked products.

Salt content. In general does not exceed 2·5 g per 100 g of product, but is usually closer to 2 g per 100 g. Salmon, sea trout and eels may contain more, up to 5 g per 100 g product.

For those products such as those prepared from salmon, trout and mackerel, for which there may be a small potential hazard from the presence of *Clostridium botulinum*, a minimum concentration of 3 g salt per 100 g of tissue fluid is recommended. But note that the salt content of the tissue itself will not only be less but will also vary according to the fat content. The relevant figures for the example of 3 g salt per 100 g of tissue fluid are as follows:

Fat content %	Salt content (g salt per 100 g product)
0	2·5
10	2·2
20	1·9
30	1·6

Water content. This parameter can also vary widely depending on the fat content of the original raw material and the degree of drying which occurs during smoking. Many, if not most, UK smoked fish products are brined before smoking. This can result in increases in weight of up to 7 or 8%, depending on the product. Some water is physically retained such that even a weight loss, due to evaporation, in excess of 10%, results only a 5% net loss in weight of the unsmoked product. Thus the water content of the flesh may be reduced by only 2 or 3% in many cases.

Smoke content. This varies widely, even for products of the same type. Total 'phenols' are likely to range up to 5 mg per 100 g of product. Similarly up to 5 mg formaldehyde per 100 g product may be present.

Compositions can vary widely. Figures given should be used with caution.

TABLE 11

WEST AFRICA: SMOKED AND DRIED FISHERY PRODUCTS

Product	Species	Brining	Smoking and heating	Composition (%)			
				Salt	Fat	Water	Ash
Hot smoked bonga (The Gambia)	*Ethmalosa fimbriata*	None	One smoking of about 2–4 h per day for 5 days Fish stacked belly down initially, then alternately belly up and down on racks over open fires in traditional smokehouse. Temperatures usually 60–80°C, but can be as high as 137°C Heavily smoked	Unsalted	No data, but in medium range apparently	18·5	
Guedj (Sengegal and The Gambia)	Various demersal species, catfish, shark, rays, large mackerel	None	Not smoked. Fish fermented (i.e. allowed to spoil) before sun-drying. Salt crystals added to drying fish to reduce fly infestation		No data		
Ketiakh (Senegal)	*Sardinella* spp. *Ethmalosa* spp.	None	Fish arranged on ground, firewood put over fish and fired. Allowed to burn out overnight, temperatures and times uncontrolled. Skin removed the following morning and fish smoked with the addition of salt crystals to reduce fly infestation Fish cooked rather than smoked		8·0% dry basis	Two samples analysed (a) 19·8% (b) 53·8% for rapid distribution	8·4

TABLE 12

ZAMBIA: SMOKED AND DRIED FISHERY PRODUCTS

Product	Species	Salting	Drying and smoking
Sun-dried			
Bream	*Tilapia* spp. *Serranachromis* spp. *Haplochromis* spp.	None	Split fish are sun-dried for 3 days, sometimes on raised drying racks, or longer depending upon the weather
Barbel (catfish)	*Labeo* spp. *Barbus* spp. *Clarius* spp. *Heterobranchus* spp. *Bagrus* spp.	None	
Karpenta	*Stolothrissa* spp. *Limnothrissa* spp.	None	Whole fish sun-dried on beach for 1 day immediately after landing
Sun-dried and lightly smoked			
Various fish	*Labeo* spp. *Barbus* spp. *Clarius* spp. *Heterobranchus* spp. *Bagrus* spp.	None	Large fish are often briefly hot smoked after sun-drying (5–10 min at > 100°C) to give a brown colour and a slightly smoky flavour to the product
Sun-dried and smoked			
Bream and barbel	*Tilapia* spp. *Serranachromis* spp. *Haplochromis* spp. *Labeo* spp. *Barbus* spp. *Clarius* spp. *Hetrobranchus* spp. *Bagrus* spp.	None	Bream and small barbel are gutted and scaled before being sun-dried for 1–2 days, sometimes on raised drying racks. They are then hot smoked (> 100°C) for 1–2 days, usually on tradtional smoking tables. Large barbel are split, sun-dried for 2 days and hot smoked for 3–4 days In some fisheries, small bream are gutted and scaled, sun-dried for 30 min to remove surface moisture and hot smoked for 1 day. This is followed by a further 1–2 days' sun-drying
Smoked fish			
Bream	*Tilapia* spp. *Serranachromis* spp. *Haplochromis* spp.	None	Scaled, gutted fish are sun-dried for 30 min and hot smoked (> 100°C) for up to 3 days depending on size

Various small fish	All very small fish	None	Whole, ungutted fish are hot smoked for 3–4 days until dry and dark brown
Barbel	*Labeo* spp. *Barbus* spp. *Clarius* spp. *Heterobranchus* spp. *Bagrus* spp.	None	Large fish are gutted, bent back on themselves so that tail lies alongside the head and hard silicate spine of the pectoral fin is speared through the tail so that fish remain folded. They may be sun-dried for 30 min before being hot smoked for 1 week until they are black and dry. Smaller fish are often not gutted. Very large fish are split, gutted and cut into pieces rather than folded
Large bream, barbel and Nile perch	*Tilapia* spp. *Serranachromis* spp. *Haplochromis* spp. *Labeo* spp. *Barbus* spp. *Clarius* spp. *Heterobranchus* spp. *Bagrus* spp. *Lates niloticus*	None	On Lake Tanganyika the fish are scaled, split and gutted. Large fish are cut into 30–40 cm long pieces. The guts are then rubbed over the cut surface and deposit a layer of blood and fat to give the end product a dark brown colour. The fish are sun-dried for 1–2 h and smoked for about a week until dry
Dried salted			
Large barbel and bream	*Labeo* spp. *Barbus* spp. *Clarius* spp. *Heterobranchus* spp. *Bagrus* spp. *Tilapia* spp. *Serranachromis* spp. *Haplochromis* spp.	Layers of split fish and salt in 44 gallon oil drum are covered and weighted down. The fish may be left for only 3–4 h or for 3–4 days; in the latter case a pickle will have formed	Lightly salted (3–4 h) fish are sun-dried for 4–5 days and heavily salted fish (2–4 days) for 1–2 days
Large bream	*Tilapia* spp. *Serranachromis* spp. *Haplochromis* spp.	Split fish are immersed in 20–30 % brine for $1\frac{1}{2}$ h in 44 gallon oil drum	Fish are sun-dried for 3 days on raised drying racks followed by 5–10 min hot smoking

Note: there are regional differences in processing. Salt is not widely used. Heavily salted fish is unacceptable and is usually exported but in some areas lightly salted (i.e. brined) fish may be bought by Zambian consumers.

Index